The real science behind the undead

Doctor Austin

ZITS BSz MSz DPep

Zombie Institute for Theoretical Studies (ZITS)

Dear Sam,

Your my top marksman student!

Doctor Justin

Cover Art & Illustrations by Russell McGovern

Disclaimer:

Whilst we endeavor to ensure our science is as accurate as is humanly and zombiely possible we do make mistakes. Here at the Zombie Institute for Theoretical Studies we love to make mistakes. They are essential to help us learn. If you spot a mistake in our textbook please let us know. We can flag up any errors online in our ZomBlog and keep them in mind when creating future editions. Don't be afraid to tell us we're wrong, we'll probably agree with you.

Although based on real science this textbook is ultimately a work of fiction. Names, characters, places and incidents are the product of fiction or are used fictitiously. Any resemblance to actual events places or people living or undead is purely coincidental.

Finally none of the descriptions of anti-zombie weaponry should ever actually be attempted by any person in any place at any time. The author takes no responsibility of any kind for the results of doing so.

Contents

FOREWORD AND COURSE INTRODUCTION

Introduction & explanation of Zombie Science 1Z

Dear Zombiology Student,

Thank you for choosing to study with us at the Zombie Institute for Theoretical Studies (ZITS) – the world's leading1 Zombieism research facility.

My name is Doctor Austin, Theoretical Zombiologist and Head of the ZITS based at the University of Glasgow's historic west end campus.

I have put together this textbook as an official accompaniment for the academic course Zombie Science 1Z. However it has been written in such as way as to hopefully make it accessible to the non-scientist with an interest in zombies. It includes standard formal explanations, extracts from my personal research journals and selections of essays I have written on the subject.

I would like to begin by saying that there are some people out there who believe that researching Zombieism is without merit or purpose. But if conflicts like those in Iraq have taught us anything it's that we should prepare for and even act upon threats to our country even if those threats never have or will exist.

The very fact there has never been a recorded account of the Zombieism condition appearing in humans makes it statistically more likely one will occur today than it would have yesterday but is not as probable as the likelihood one will occur tomorrow. By taking this course you have taken your first step into what is perhaps one of science's last unknown frontiers.

I would also like to thank all the scientists who have ever existed. Every breakthrough made starts a never ending domino rally of

---

1 At time of print the Zombie Institute for Theoretical Studies is the world's only Zombieism research facility.

discovery. We all form part of the science chain and by stealing, ahem, sharing ideas we all get closer to the answers we seek.

At the end of this text is a bibliography of sources. These are some of the definitive texts on the subject of zombies and surrounding areas and we recommend you buy every last one of them.

Best of luck,

Doctor Austin ZITS BSz MSz DPep

A word from the Minister of Zombies:

I first became aware of Doctor Austin's work when I witnessed one of his lectures at the 24th World International Zombiology Zymposium (WIZZ) in Blackpool during the late 1980s. Sharing the stage with some of the world's most influential experts on the walking dead, I was sceptical to say the least. His obvious youth, those geeky NHS glasses and a virulent green polyester shirt all played well to the fashion-conscience audience but it was not until he started to share his research that the whole hall sat up and listened.Much of the research he outlined during his lecture was radical, even heretical for the established members of the zombie fighting community. At one point, a whole clutch of those who see witchcraft as the main course of Zombieism, got up and left, leaving a trail of twigs and leaves behind them. There was more than one spell cast against the young pretender that day I can tell you.

Still, those of us on the scientific side the zombie origin debate listened intently…

Many years on and after a very successful tenure as both a Special Scientific Advisor to the Ministry of Zombies and then in his current teaching role at the ZITS, Doctor Austin, has finally published the core of his research in a modular training course for the scientifically challenged. Within these pages, you will find invaluable information across the full spectrum of Zombiology in a readable format which everyone can understand. This talented academic seems to have taken it upon himself to educate the world on the true science behind the walking dead and his qualification is now a mandatory requirement for any new recruits at the Ministry of Zombies. I fully expect other zombie organisations to follow suit.

So, reader what are you to expect from this genius of the Zombic world? Well, there was a time when all a zombie fighter needed to do was grab cricket or baseball bat and stay ready for the apocalypse. But, times have changed and you'll find within this volume dozens of modules covering everything from the origin debate to the role of hand washing in any upcoming war against the walking dead - yes, cleanliness is always high on the agenda with Doctor Austin.

Those individuals kept awake by that tapping on the window or fearful of the rotting claws of the undead reaching to get them in the shower, need to study this book and complete the online examination. Better still make sure you catch one of the public lectures, of which the Ministry of Zombies is but one proud sponsor.

Now, it's time to take your first step into the weird and twisted scientific world that is Zombie Science 1Z.....be brave and let the geek in the glasses be your guide...meet at you at first break by the drinking fountain.

Sean T Page

UK Minister of Zombies

Author of the Official Zombie Handbook UK & War against the Walking Dead

Profile – The Zombie Institute for Theoretical Studies (ZITS)

Located within the main campus at the University of Glasgow the Institute is a historic building with a colourful past. It first opened in the early 19th century when public fear of Zombieism was at an all time high and citizens demanded answers from the learned community. At this time it found 1% of all zombie's seen rising from the dead were in fact just misdiagnosis of death. The other 99% were intoxicated Glaswegians2. This figure fell drastically after the introduction of improved practice amongst medical professionals. Not unused to being at the centre of controversy the Institute came under fire (physically as well as metaphorically) in 1856 when eminent scientist Lord Kelvin fired his blunderbuss through the front doors. The Institute had recently definitively decided to call the collective symptoms attributed to a zombie Zombieism. Lord Kelvin strongly felt it should be named Kevlinism as at that time every discovery, place and thing in Scotland was named after him. Fortunately the heat shifted from ZITS after Kelvin publically announced that heavier than air fly ing machines were impossible and X-rays would prove to be a hoax. In the mid 1930s when the Institute purchased its first aeroplane they named it the Kelvin in his honour.

Staffed by the most experienced Zombieism experts of their field the Institute also boasts an extensive range of research facilities including; SETI3 based tracking systems that monitor the skies for extraterrestrial Zombieism threats, equipment capable of DNA sequencing and analysing genetic samples at high speed and a PC with access to Wikipedia.

The institute has seen many of science's great minds pass through its doors. Notable graduates include: Dr E. Brown, Professor B. Campbell, Sir Dr Professor G. Romero and Dr P. Venkman.

---

[2] Glaswegians are natives of Glasgow a settlement located on the Clyde River in Scotland, UK. They are famed for a natural disposition toward imbibing alcohol for more information research Connolly, Billy.

[3] Search for Extra-terrestrial Intelligence.

In 2010 ZITS received its first ever public engagement grant from the Wellcome Trust charity. This allowed them to develop a public lecture series led by Doctor Austin that toured venues across the United Kingdom during 2011.

The Institute continues its research to this day and is always on standby for zombie sighting and outbreaks. It also runs the Zombie Science 1Z degree course amongst others in the field of Zombieism. Please feel free to drop by the Institute, public engagement and science communication is also a part of their brief.

Staff Profile – Doctor Austin
Doctor Austin ZITS BSz MSz DPep

With a worldwide reputation as the leading expert in Zombieism Doctor Austin has been a Theoretical Zombiologist for more than sixty years. He holds a Masters in Zombiology, obtained from the ZITS, a Bachelors of Zombotany, whilst studying in America at Eerie Indiana State University, and a Doctorate of Pepper from the renowned University of Peterhead.

He was first led to this field during the early 1960s whilst studying the prion disease Kuru in Papa New Guinea. Here he saw his first glimpse of a then unknown infectious agent that was capable of producing some of the symptoms found in Zombieism. Kuru was spread primarily by the ingestion of infected brain tissue during cannibalistic feasts amongst local people. This at first tenuous connection between Kuru and the fictional zombie's famed love of devouring brains led Doctor Austin down a path that would see him formulate a theory implicating a rogue prion as the most probable cause of Zombieism.

These ideas were strengthened in the late nineteen eighties and throughout the nineties when Doctor Austin was called in to aid in the bovine spongiform encephalopathy (BSE) outbreaks amongst British cattle and the subsequent instances of the human form of mad cow disease, variant Creutz-Feldt Jakob disease (vCJD).

In 1993 Doctor Austin accepted the position as Head of the Zombie Institute for Theoretical Studies. He is the seventy sixth fellow to hold the office and received his induction from out-going retiree Professor B. Campbell.

In 1996 he commissioned the Institute's largest research task to date. The 'Noah' Project was established to observe two of every animal species for signs & symptoms of the Zombieism condition. Over five thousand animals have been recorded and logged during the past fifteen years and this mammoth task, likened to the human genome project, is still ongoing.

Prior to his scientific career Doctor Austin worked on product design within the private sector. Perhaps his most notable creation was the Ultra-Safe Concrete Children's Swimwear, the catchy slogan, wherever you leave them, that's where they are. Doctor Austin and his invention were featured on the television series Dragon's Den. Unfortunately repeated requests to air the episode have been rejected.

In 2008 Doctor Austin received the great honour of being appointed Zombiologist Royal to Her Majesty the Queen of England. He carries out yearly inspections of the Monarch for the signs and symptoms of Zombieism.

He has also published several books on the subject of Zombiology including the Richard & Judy recommended "Zombieism in the 21st Century" & his New York Times bestselling autobiography "Please. Don't eat my face."

Course outline

Zombie Science 1Z is a twenty credit course assessed by an online examination. All students must complete the online examination in order to obtain a pass.

The online exam can be found at **www.zombiescience.co.uk**

Whilst home and online study options are available we strongly recommended attending the accompanying lecture, Zombie Science 1Z led by Doctor Austin, details of these can also be found on the website.

The textbook is set out like the course itself, in three modules.

Module one (M1) the Zombieism condition examines the symptoms associated with Zombieism. Using those observed in the fictional zombie as a starting point we can then identify which are most scientifically feasible. From there we can begin to establish an accurate picture of what a real zombie would be like and how it came to be infected.

The second module (M2) Zombieism causes uses the symptoms outlined in M1 to track the condition back to its source. Here we will gain an overview of the variety of ways diseases can come about in humans and go into detail on what we currently believe to be the most probable source of a Zombieism outbreak in the 21st century.

Our third and final module (M3) is preventing & curing Zombieism. This module is perhaps the most important with regard to public perception of the Zombieism condition and those who suffer from it (known as 'zombies'.) Counter to the suggested methods of treatment in popular culture (e.g. blunt force trauma, decapitation, combustion, etc) we put forward more humane strategies such as diagnosis, medical treatment and developing cures. Whilst more violent methods are included these are purely informative and to be used only in a last resort scenario. This module also describes what a Zombieism outbreak could actually be like.

Alongside this textbook, the lectures and lab classes' students are also advised to cast their study net as widely as possible. As we will go on to see developments in the study of Zombieism have emerged from every discipline of science and even from lengthy sessions in front of game consoles.

Introduction: What is the fictional condition of zombieism?

Hopefully you'll all have seen a zombie film, if you haven't you will likely fail this course. But you still get points for trying and you know what points mean...? Prizes no, nothing. Points mean nothing at all, this is an academic course.

Like your first kiss everybody always remembers there first zombie. Dependant on your age a different image will come to mind when we say the word, zombie. Older students may recall the original slapstick scary black and white zombies of George A Romero's now legendry Night of the Living Dead. Those of a younger generation may fear the fast moving hyper gory ghouls of computer games & blockbuster films whereas students joining us from preschool classes will be most familiar with zombie Lego.

The arguments as to what collective group of symptoms constitutes a fictional zombie would make a course of its own and it already has its own group of experts researching it daily. These people are known in the wider community as nerds.

The Zombie Institute for Theoretical Studies has defined the following description based on the writings of eminent zombie experts George A. Romero and Susan Sparrow. We take this to be the standardised definition. As we will go on to describe actual Zombieism would lack many of these symptoms or they would manifest in a different way than described. View this outline as a starting point but do not take it to heart. Media and other elements of popular culture have created a large amount of conventional wisdom regarding zombies that is wildly, even dangerously, inaccurate.

Fictional Zombie Outline:

The fictional Zombie is the reanimated corpse of a living thing primarily of the human species. It acts on a basic level with one goal - to bite another living thing. This goal works on many levels: it allows Zombieism to be passed on, it provides nourishment (or the perception of it) and it combats predators. Although technically living they are nonetheless dead and their bodies continue to decompose. This condition can effect mental functions, impair movement and whilst inhibiting the feeling of pain the body is left unable to fight off biological threats or heal. Verbal communication is restricted to an unpleasant sounding moan. (Romero, 1979)

It is clear from this wide range of symptoms that a Zombieism condition would have to be extremely complex. Let's discover more about it as we begin with module one the Zombieism Condition.

MODULE ONE – THE ZOMBIEISM CONDITION (M1)

Overview of typical Zombieism symptoms and examination of if/how these symptoms are being caused

A scientist first encounters a new medical condition usually when a patient comes in complaining of certain symptoms. By identifying the symptoms we can then make an educated guess as to their cause. As Zombiology students you must be able to instantly recognise all the symptoms associated with Zombieism. Below is our aforementioned list of traditional Zombieism symptoms as found in human patients in film & media.

- Dying then coming back to life (referred to as being 'undead')

- The continued decomposition of skin and organs (as per normal human decomposition)

- Poor balance and coordination leading to a shuffling unsteady walk

- The loss of reasoning, memories and other human personality aspects

- An unending desire to consume living flesh with a particular fondness for the brain

- An apparent inability to feel pain or heal

- Communicating only with an incessant moan

Doctor Austin states, "I must insist you find more than one of these symptoms before making a positive identification. For example all of my ex-wives emitted an incessant moan, but not all of them were actually zombies."

It is worth now taking a leaf from the nitpicking nerd and asking that all important question. Which of these symptoms are scientifically unlikely? By answering this we can disregard them from our studies and start to compile a more accurate picture of a true zombie.

Please take a moment to consider the symptom list and formulate your own ideas on what seems implausible then go on to read the remainder of this section.

The first symptom and perhaps the most controversial is a zombie's ability to go from corpse to creature of the night in a very short time.

Dying then coming back to life

A Zombie trademark has always that they are undead, a deceased thing that behaves as if it were alive. This doesn't seem to make logical sense. As we know the brain acts as the command centre of the body. The area called the medulla is responsible for sending electrical impulses that control our heartbeat and respiration. Once we die the heart stops beating, the lungs cease respirating and the brain switches off leaving the body unable to function in any way. Although science still has a lot of unanswered questions regarding the brain the concept that it needs to be switched on is universally agreed. The brain itself dies after three to seven minutes without oxygen.

Doctor Austin's Journal 20/04/1998

Location: Adams Mortician, Guerniville CA USA

Time: Just after lunch

I have arrived in the former British colony of America to investigate a potential instance of human Zombieism. A woman was observed staggering around mumbling incoherently and biting people. After establishing for certain it was not another Mel Gibson4 I attended immediately.

---

[4] A Mel Gibson is the ZITS phrase for a false call out caused by an inebriated person acting like a zombie. It has the name Mel Gibson because he is the cause in over 75% of cases.

Unfortunately the patient had been hit by a vehicle known as a pickup truck and was now deceased. Upon further enquiry I was told it had been an accident. I asked why the body show signs of being ran over multiple times, as if the vehicle has driven forward and backwards over it. They told me it had been a bad accident. I facilitated access to the body and secured her brain for an extra $4.95. Sometimes working in America has its advantages.

Without being able to observe the patient in her natural state it was hard to see any of the symptoms. I removed the brain using my tin opener5. I gave it a visual check, took tissue samples and conducted a short series of Krang6 impressions. All this led me to the same conclusion, I had in my hands, a human brain.

I then decided to use the opportunity to test out the zombie undead theory. After replacing the brain I sat back and waited. Perhaps my current research would be proven wrong and this woman would return from the dead.

Doctor Austin's Journal 24/08/1998

Location: Adams Morticians, Guerniville CA USA

Time: After breakfast (but before brunch)

More than four months have passed. I have lost over half my body weight vomiting from the stench of this woman's decomposing corpse. I honestly think she might not have Zombieism but I must stay focused. I'm sure there were times when OJ's lawyers occasionally doubted him. Furthermore the motel owners are not being very understanding about the experiments I'm carrying out in my room. People keep complaining about there never being any ice in the machine.

Doctor Austin's Journal 31/10/1998

Location: Adams Morticians, Guerniville CA USA

Time: My watch is broken

---

[5] Tin Opener is a brand name surgical tool, not an actual tin opener, they are rather inadequate for human brain removal (except in emergencies)

[6] The evil brain from the Teenage Mutant Ninja Turtles

A turning point, the local sheriff informed me the dead woman has been identified as a lady of the night. I rather angrily asked him why he allowed ladies out during the night when they should be encouraged to come out during the day. He became even angrier when I asked for the corpse 'to go' and requested I return the corpse from the motel then leave the state.

I might not have found Zombieism but I found more evidence against the idea that humans can return from the dead.

<Entry ends>

If a zombie were brain dead it would be not just incapable of moving but incapable of anything. Realistically this would make for a somewhat dull movie, a band of humans fighting against lifeless corpses, one sided certainly. The clue to this Hollywood hypocrisy is of course in the name. Undead. Precisely a living thing that was not alive and was therefore dead transferred to state whereby it has become alive or in other words the reverse of dead giving us, undead.

This can also provide some relief regarding the odds of one being pursued by a severed zombie appendage, a routine occurrence on the silver screen. In most animals, including humans, motor functions are controlled by the brain. It is the command centre sending signals down the spine. There is no replacement or substitute part within the human body. Without creating new motor control areas (brains) within the severed appendages there is no possibility of them becoming 'alive' and independent of the body.

Following this line of reasoning it should be noted that if a fictional zombie's head is severed from its body and the brain is not destroyed the head may still pose a risk. Tests at the Institute are ongoing regarding this but for the moment it seems if fictional zombies somehow came into existence you can feel free to lop off body parts safe in the knowledge they won't return to brutally murder you later or feel up your girlfriend.

To have now reached the opinion we are all technically 'undead' indicates you are beginning to think along the right lines for this course. However a stipulation may be that you have to be alive then legally dead for a brief time before returning to life in order to qualify for the full undead label.

Ultimately it is not possible (at the moment anyway) for a zombie to be a deceased thing that behaves as if alive. When a Doctor resuscitates a dead patient with a defibrillator they do not enter a new state of biological existence known as undead they are just plain old alive.

Leading on from this deceased idea these fictional zombies face should face a whole host of other problems if their bodies truly were dead and began to decompose.

The continued decomposition of skin and organs

Decomposition is name given to the biological and chemical changes that occur soon after death. How soon, well approximately four minutes after the death of a human decomposition starts to take hold. The process of decomposition is a continual one in which organic tissues and structures progressively decay and become disorganised. It is generally divided into five stages, in some cases not all or even additional stages may be present. These five stages are Fresh, Bloat, Active and Advanced Decay, and Dry/Remains. If a method was developed by which human beings could be reanimated after death, as occurs widely in fictional Zombieism, these five stages might act as the life cycle of such a zombie.

Stage 1 – Fresh: (0-3 days after death)

Once our hearts cease to beat blood begins to settle in the most dependant parts of the body. Blood, reacting to the effects of gravity, also collects in areas closest to the ground. Minutes to hours following death comes livor mortis causing parts of the body to appear discoloured at vicinities where blood is accumulating. Indeed these are known in the industry to embalmers as post mortem stain. These red-purple stains are capable of moving around the body if the corpses' position is changed because the blood is unfixed.

A fictional zombie at this point would look something like a mood ring. Changing colour as its movements forced remaining blood around its body. It would only have this ability for a maximum of 8 – 12 hours. Then livor mortis would become fully onset, blood would become fixed and the skin would lose all muscular control. Skin could contort creating a disfigured misshapen appearance and bones would become very prominent.

If this weren't bad enough rigor mortis would also being going on alongside. As the body relaxes in an initial bout of primary flaccidity briefly upon death, just long enough to deposit ones bowels, muscles then start to stiffen. It starts with the eyelids, neck and jaw, hence why closing someone's eyes is done promptly upon their death. The overall effects of rigor mortis are variable depending on other factors. If the body had been undergoing strenuous physical activity prior to death it would set in far more quickly. This casts some light as to why wealthy, aged bachelors arrive at the morgue still smiling with a wilfully erect penis after dying during the throws of passion with their twenty something mistress.

Instances of post mortem erect penis are also high in cases where the victim has been hung or suffered a violent death. Working in partnership with the potential for accumulation of blood due to livor mortis a body that is vertical might also feature an erect penis.

Instances of zombies with raging hard on's should technically be much higher in fictional zombie portrayals than is currently seen. Particularly if we take into account that they are nearly always standing up, initially at least, therefore the any remaining blood is guaranteed to be flowing downwards.

A cold environment will also facilitate a rapid onset of rigor mortis. The deceased's age, weight, physical condition and muscular build must also be considered. When dealing with baby or child zombies the rigor mortis would be somewhat less pronounced due to lower concentrations of muscular tissues. They would certainly be a more limber foe an added advantage alongside the moral implications of popping open a tiny zombie baby's head like a melon.

We then reach the final of the mortis trilogy with algor mortis. The body begins to cool down and match the surrounding temperature. Muscles begin to get flaccid one again. All this is occurring immediately after death for about 24 – 48 hours. It is usually within this time that a human body has undergone some kind of treatment by a medical professional or mortician.

The anus and vaginal areas would be stuffed to prevent leakage. The eyes and mouth may be glued shut. Limbs are injected with an embalming fluid that stiffens them up and allows them to be positioned. A mortician is trying to make a body look as 'human' as possible for the sake of loved ones.

A fictional zombie that was reanimated from the grave or morgue would have this to contend with. It would have great difficulty getting its eyes open or its mouth to utter its trademark moan. It would fortunately look rather handsome, assuming it was handsome before death.

If a body had not been found after death and remained open to the elements it would then enter stage two. A zombie fortunate enough to have been buried first will decompose eight times slower than one exposed to air.

Stage Two – Bloat: (4 – 10 days after death)

An important process throughout decomposition and particularly in the bloat stage is autolysis also known as self digestion. As the body can no longer use oxygen to help purge it of nasty chemicals they begin to accumulate in our cells. Eventually they poison our cells from the inside out and they decay and die. It takes a few days for this process to become visually apparent.

Initially blisters appear followed by green patches accompanied with foul smelling odours. This is a good indication we've entered the putrefaction part of decomposition. Putrefaction is the destruction of soft tissues by micro-organisms such as bacteria. These bacteria emit gases which bloat up dead bodies. The body then begins to heat up again. A chemical reaction often observed when opening an old bag of rubbish. After a few weeks of swelling something very disturbing and very messy will occur. Some zombies could explode. Although this is likely to be the exception rather than the rule, all of these fictional zombies should be bloating up in some respect. Furthermore these gases can cause the eyes and tongue to protrude, push stinking blood stained fluid from orifices and force the intestines out via the rectum and/or vagina. The gas itself is also exuded, primarily from the anus, and can be forceful enough to tear the skin. You don't see this portrayed in fictional zombie cases. Perhaps Hollywood found the truth to be worse than fiction in this instance.

As always we must consider the environmental factors. In a dry heat like the Sahara dead zombies could mummify. With no way to replenish their cells with water they will stumble around slower and slower until eventually they become no more than an extra in the next Indiana Jones travesty.

If you are studying this course in Britain, or hail from the country of Scotland where our Institute is based, a concern may now be that you live in an environment where it is extremely unlikely you will ever see the sun.

This need not be a concern. It is unlikely a dead zombie could cope with the cold either. A decomposing zombie is dead meat. We encounter dead meat frequently, fish fingers, hamburgers, Walt Disney's severed head it is all virtually the same. After about seven days you have to dispose of it when it starts to turn. A logical assumption could be that cold is in fact a friend to dead meat, after all freezing meat products lets you keep them for long periods. But unregulated cold really messes with living systems. The human body is made up of a large percentage of water, varying from person to person. When water gets cold enough it freezes. Ice crystals then form and expand. These crystals can break easily through brittle, frozen human tissue. Frozen, fictional zombies in a location such as the Antarctic could, unlike the T-1000 in Terminator 2, easily be shattered into pieces should the mood grab you.

Although fictional zombies may be lucky enough to enjoy some level of preservation should they then begin to thaw, the normal decomposition cycle will kick back in. At that juncture freezer burn, nemesis of the ice cream as well as apparently the zombie world would also be a major concern. As these fictional zombies are chilled during the night and subsequently thawed out again during the day freezer burn would rapidly set in.

It is also during this stage, in a more balanced climate, that insects will start to pose a problem. Despite all the advances in chemical and technological warfare devised to stop them they are still troublesome beasties to us people. Even more so in exotic countries, where in Zimbabwe a fly the size of a small car organised a successful coupe de tat and became President for seven months. Fortunately our immune systems, as well our hands, are excellent at fighting them off. In fact it's all that's really preventing us from having our tongues and eyes consumed by flies and maggots within seconds.

A Hollywood style dead, decomposing zombie would be home, sweet, home for a range of insect life and any zombie in a region prone to flies would quickly become infested and find himself the feature of a Sir David Attenborough Life documentary. Certain flies would begin to appear in the first few hours but it takes several days for them to settle in and lay eggs. As they begin to hatch soft tissues would be devoured and a zombies eyes (if not already glued shut) would stop working rather soon. These are only insects. Dead zombies could become prey to all kinds of wild (i.e. bears, lions, and giraffe's) and domestic animals (i.e. dogs, snakes, hamsters).

Stage 3 – Active Decay: (10 – 20 days after death)

Once the gases have escaped the body by any means necessary and assuming our zombie has managed to survive this far it will be rewarded with the elite, black putrefaction. It will receive a catwalk model figure or to be more precise it will lose a large amount of its mass, mainly due to it being devoured by insects who can now more easily gain access through blown open gateways. Certain chemical reactions that turn fat into soap will give skin a yellowing waxy like coating in some instances. This effect is less common amongst slim built people because fat is necessary for the effect. We can therefore assume it will be prominently seen during a zombie outbreak because we all know there will be no overweight people after a zombie outbreak. This isn't said simply for random cruelty. It is just pure physics. An overweight person can no more out run a member of the living dead anymore than they can prevent there mass from exerting a gravitational pull.

If at this point in its decay a zombie is unable to move it may be lucky enough to gain enviable an status symbol amongst zombies by procuring a CDI. A CDI is a cadaver decomposition island. This is the area of dead vegetation that forms around a body as a result of those nasty fluids that leak out.

Zombies would then virtually disintegrate, large section of epidermis would peel off in a process known as skin slip. They would then start to dissolve as well as becoming even stronger smelling. It is interesting to note that two of the most prominent gases in decomposition are putrescine and cadaverine, really quite apt. Cadaverine is actually not just associated with putrefaction. This diamine is produced in very small amounts in urine and semen. It is believed to help give them their distinctive fragrance. Likewise with Putrescine, it too is found in semen and may contribute to bad breath as well. Eventually the body will become so toxic that even insect pupae will flee from it to be born safely. This indicates progression to the next stage.

Stage 4 – Advanced Decay: (20 – 50 days after death)

It will be a relief to know that the stinky smell will thankfully diminish somewhat during this stage and insect activity will finally curtail, primarily due to there being very little left for them to eat. If not already the case these zombies would be completely blind making them easier than ever to evade. Insect activity has not altogether ceased, far from it, in fact at this phase there may be the greatest insect diversity. Beetles, ants, flies, mites, a regular Pixar animation cast are present. On a side note the study of which insects are present during decomposition was essential in allowing scientists to develop the CSI brand of television programmes.

Stage 5 – Dry/Remains: (50 – 365 days after death)

All that has become of our zombie is some dried skin, hair, cartilage and bone, the area's most resistant to decomposition. It is now visually more a monster from Jason and the Argonauts than Dawn of the Dead. However the brain itself will be long gone making it further unlikely a zombie could be in any way active by this point. If all soft tissues are gone it can be described as skeletonised if not then only partially skeletonised. Bones themselves can take months, years, decades, even centuries to break down into dust or fossilise. Fossils are the best evidence of this, obviously not if you are a creationist. This is the longest of all stages of decomposition.

As reiterated throughout these stages the exact timeframe is heavily dependent on individual circumstance. Many things can alter the process of decomposition. If the body could be protected from factors like insects it would make a difference. Spraying these fictional zombies with insect repellent might certainly make them last longer. In a murder case in America a deceased woman was found and thought to have been recently killed. She had actually been dead for over four months. The killer's use of insect repellent and other chemicals had sterilized the body from insects and vastly slowed decomposition, even internally.

Although the fictional zombie life cycle runs to as much as 365 days it is highly unlikely a zombie would last that long. Becoming blind within a matter of days and rapidly losing muscular control would be its main concerns.

It is plainly seen that just one of the above described negative implications of decomposition could halt a zombie before it starts. The point is a zombie would have to endure a series of them. The most threatening decomposing zombies would be those who were preserved greatest prior to turning. Statistically this makes Vladimir Lenin and Evita Peron the most dangerous zombies and Egypt the country that would find the highest concentrations. A safe distance should also be kept from any museums with a 'Wonders of the Pharaohs' section and peat bogs. Those working off world should also consider that the vacuum of space and very low temperatures would virtually cease the process of decomposition. If you are an astronaut and encounter a zombie in space you should attempt to push it in the direction of the sun.

The Institute has also done extensive studies into whether necrosis might be part of the Zombieism Condition. We're all shedding cells faster than Katie Price sheds blokes, around a 50 - 70 billion are replaced every day. The normal process of cells dying is known as Apoptosis. Certain circumstances, such as infections, cancer and spider bites amongst other things, can bring about the condition Necrosis in our cells.

Necrosis causes our cells to break down in a disorganized manner. It is very different from what happens to our bodies when they normally die.

It is worth taking the time to track down and examine a picture of someone who has this condition affecting a limb, perhaps in the instance of a spider bite from the Brown Recluse Spider on the leg. You will see the leg most certainly looks as if it'd waltz through any zombie audition. Actually it wouldn't. Commonly the leg is amputated from above the knee.

The point is that if Necrosis was part of the Zombieism Condition and went untreated any zombie would soon virtually disintegrate just as a decomposing one would. Luckily unlike decomposition Necrosis can be treated.

In conclusion if a human body infected with Zombieism actually decomposed like a normal body, whether by normal processes or the involvement of Necrosis, they would barely last long enough to gather into groups, bite anything or get anywhere near a shopping centre. Zombies cannot therefore be decomposing in the usual human body manner and this symptom can be eliminated from our particular direction of study.

An unending desire to consume living flesh & brains

The staple diet of the fictional zombie is widely known to be human flesh with a particular penchant for the brain. As will be described in more detail later in the course damage to the brain can induce an uncontrollable urge to eat. But when it comes to cannibalism it is difficult to discover a recognisable medical cause to explain it in humans.

It is certainly a difficult area to study. Academics and scientists constantly disagree regarding cannibalism's role in human history. Often cases were reported by white explorers encountering foreign tribes for the first time. They used terrifying tales of cannibalism amongst these odd new fellows to incite fear from those back home. The fear would feed the funds of these explorers to continue to bring back more juicy gossip. Bear in mind there was no internet back then.

Furthermore a great taboo surrounds cannibalism in western culture. So much so that there exists little laws relating to the act itself, those found to have committed cannibalism are usually convicted for the act of murder rather than the consumption itself.

Cannibalism has existed in various forms throughout human history. Survival cannibalism occurs when other people are the only available food source in a desperate situation. There are many documented occurrences, some have even become films. This is the only forgivable type of cannibalism available in the west.

Endocannibalism involves consuming only members of one's own family. This was practised amongst the Fore tribe of Papa New Guinea up until the 1950s & 60s. It was part of their religious beliefs and indeed the belief's of many early human cultures. The idea was that the consumed person lived on, along with their wisdom and power, within the consumer.

<Entry Begins>

Doctor Austin's Journal 01/04/1968

Location: Waisa, Papua New Guinea Highlands

Time: Shortly before sunset (they don't use watches here)

My expedition has had a bad start. This morning I used my tie to mimic a snake and give my guide Suna a comical scare. He reacted quite badly. I kept shouting 'April Fool's' at him in tok pisin, but my pronunciation isn't the best, I think I might have simply been loudly calling him a female idiot. An American Doctor later told me they don't have April fool's day here and that my tie is the same colour as an extremely deadly snake.

With the snake tie incident behind us we headed out to a more remote hamlet to meet an elderly man. Suna assures me he can tell me something of the cannibalistic feasts that once took place here.

The place was deserted when we arrived, a typical situation when they knew I was en route. Unfortunately my name, Doctor Austin, means White Devil in almost all of their local dialects. Fortunately the old man was suffering from Kuru and unable to walk. This meant he was still there when I arrived. Suna translated.

Me: Have you ever taken part in a funeral feast?

Old Man: You kill me?

Me: Look angrily at Suna No, have you, ever eaten another Fore person?

I show him a picture of Anthony Hopkins as Hannibal Lecter for clarity, he looks more confused

Old Man: There were many mumu (feasts) when I was a boy

Me: Did you take part?

Old Man: I often received pieces of the brain

Me: What did it taste like?

Old Man: Kakaruk (chicken)

<Entry Ends>

It was usually the women and children who consumed the flesh normally because the adult males tended to die first.
Exocannibalism, or the eating of others, has three potential manifestations. It could be a means of inciting fear amongst one's enemies, an attempt to steal their life force or simply a need for nourishment. Perhaps the most unusual case in the 20th century involved a Japanese man named Issei Sagawa. In the early 1980s he murdered a Dutch student and ate parts of her body. By eating her he sought to acquire some of her health and beauty, qualities he felt he lacked. He described the meat as being soft and without much scent, a little like tuna. Sagawa was arrested by the French authorities and held for two years before being extradited to Japan. A mental hospital found him sane but evil and he checked himself out in 1986. Luckily he managed to carve out a life for himself as a minor celebrity and food critic. This clearly indicates cannibalism can sometimes have a happy ending.

Zombies in fiction tend to fall into the Exocannibalism category. They appear to consume only living humans. Because we know their brain and other internal organs are nearly inoperable it does not make sense they are consuming to stem hunger. However the fact they move at all proves they must need energy of some kind so they must be getting some nutrients through consumption.

But what is it that is producing this selective cannibalistic behaviour. It is interesting that in popular culture zombies never bite each other, even by accident. Given that they will be partially or completely blind this is quite amazing. Let's look at an instance of this behaviour in a non-zombie.

Armin Meiwes is a German Cannibal of some note. In 2001 using internet chat sites he scouted out a volunteer to be killed and eaten. Bernd-Jurgen was his man and the pair met up in March of that year for a meal. The starter was Bernd-Jurgen's penis, initially Meiwes tried to bite it off, when that failed a knife was used. He tried some raw but it was too chewy and so set it on fire. Afterward the penis was so badly burned Armind cut it up and fed it to his dog. The evening was then soured somewhat as Bernd-Jurgen realised he was the main course. Armin stabbed him in the bath and went on to eat around 20 kg (44 pounds) of the body over a period of around 10 months.

This was a difficult case to judge as both men were willing participants. Armin described his motive as being related to a desire to have a brother, someone to be part of him. Throughout his defence he argued it had been a mutual pact rooted in a sadomasochistic homosexual fantasy. Without a specific law for cannibalism Germany convicted him of manslaughter, and he received eight and a half years. However this was overturned in 2006 and he was instead convicted of murder and sentenced to life imprisonment.

Most recently Stephen Griffiths nicknamed the Crossbow Cannibal confessed to multiple killings in the UK. He claimed to have skinned his victims cut off multiple parts then cooked and eaten them. A failed effort at baking was foiled by a faulty cooker. He also ate sections raw in a flat he nicknamed the slaughterhouse. When arrested and interviewed by Police he stated the cannibalism was "part of the magic". Police have found no evidence of this aspect to his crimes and many believe his claims are merely attempts to increase his notoriety.

All of these stories certainly support the image of the fictional zombie, a creature mercilessly chomping on raw limbs. But this behaviour has to be brought about as a result of the Zombieism condition. All zombies exhibit it and it is unlikely they were all into Vorarephilia7 before they turned.

Seeking a prospective cause from our cannibal killers is difficult. In all three of our cases other factors were always present. They all derived some manner of personal or sexual pleasure. Whilst a zombie may moan frequently it is not in orgasm as it devours a victim. With Griffiths he sought the power and notoriety of the serial killers he studied. Having all these potential causes and reasoning behind cannibalism makes it hard to attribute one specific reason for it to exist in all zombies.

If we examine the medical opinions given for these cannibal killers it might help us discover a more root cause of cannibalism. Mr Sagawa our Japanese cannibal was examined upon his return home by psychologists at Matsuzawa Hospital. After extensive study they determined him to be evil but not insane. There was no conclusive evidence as to exactly what made him a cannibal.

At the retrial of Mr Meiwes in 2006 a psychologist described how he [Meiwes] still fantasised about eating young men's flesh and could not be allowed free. However no specific condition was ascribed to him and at no point has he pled insanity.

The self-proclaimed crossbow killer was convicted of murder and also made no attempt to claim himself mentally incapacitated. Cases certainly exist in which a cannibal killer has been found mentally unstable or insane and this has been used to explain the behaviour. If we look at the current formal index of insanity there is no mention of cannibalism.

As the zombie lacks the ability to be influenced by other factors, i.e. it cannot commit cannibalism for sexual pleasure or to incite fear, and it likely lacks enough mental capacity to suffer from a mental disorder like insanity, it becomes more and more difficult to hold up the image of the zombie as a cannibal.

But as we will go on to discover it is possible for a human to get a compulsion to continually eat. A supposition for the zombie could be that is it mentally compelled to eat any source of food it comes across. Taking into account its poor vision and lacking our standard object recognition abilities it might accidently eat many things before finding something it can digest. The fact a zombie

---

[7] Vorarephilia is a sexual fetish whereby one is aroused by the idea or act of being eaten.

sometimes bites and/or eats a human is simply a side effect of the fact humans are often the most prevalent thing in their vicinity.

Once again we never see this type of zombie eating behaviour in movies. Whilst pursuing a victim a zombie doesn't suddenly begin eating a nearby lamp. Technically, it should. Like fingerprints every zombie is different. Some may still have a slight sense of smell and others clearer vision making some meal selection possible.

Furthermore such fictional zombies should be drawn to the nearest food source, meaning a group of them should stop pursuing you if you passed a supermarket, but not McDonalds.

Therefore a better description of the zombie diet, rather than a desire to consume human flesh and brains, would be an unending desire to consume any food source.

Inability to feel pain (congenital insensitivity to pain)

So far the zombie hasn't really lived up to its fear inducing reputation. They come across as smelly, half blind, leaking nuisances with the munchies. But fictional zombies are usually attributed with one super power, an inability to feel pain. You can slice them, you can dice them, you can stick their face in a blender but even with multiple limbs removed and a face like a smoothie the zombie still carries on with its sole purpose.

Whilst it may come across as the perfect superpower in actuality it'd make a zombie far more like Mr Burns than Mr Fantastic. The central nervous system does a great job of letting you know when you are or have been damaged. Sensory neurons detect things related to our five senses and transmit them as signals to the brain. The brain translates these into something we understand. Like a feeling of pain from something sharp or a feeling of burning from something hot.

To realise just how useful this ability is think of all the times you've felt pain, paper cuts, stubbed toes, partner left you for someone more attractive even though they are empty inside compared to you, scraped knees, etc.

Without having felt those feelings you wouldn't have realised when to stop doing an activity that was hurting your body. Standing on a nail for example. A fictional zombie that feels no pain would continue in a misguided attempt to eat a spinning helicopter rotor or persist in visiting Bruce Campbell for coffee.

Those are only external injuries. Without feeling internal pain a zombie wouldn't realise if he had a sore tummy, sexually transmitted disease or tumour.

Loss of feeling in limbs can be brought about in several ways. A lack of blood supply is something we have ascribed to the fictional zombie during the decomposition process. If the zombie had been sitting in an unusual position, cross legged for instance, prior to pursuing you it might have a nasty case of pins and needles. In both cases the zombie would still be feeling something when you started savagely beating it.

There is a neurological condition that could provide the answer to this aspect of fictional Zombieism. It is called congenital insensitivity to pain and is an extremely rare condition that some people are born with. It appears to affect part of the peripheral nervous system creating an indifference to painful stimuli.

It is obviously an arduous condition to suffer from. Pain is an important teacher in our lives. Children with this condition repeatedly put themselves in harm's way, not learning from past mistakes. They continue with behaviours that are damaging to the body such as chewing and biting on their own tongue. Many don't live beyond the age of 25 because minor conditions develop unnoticed into far more serious ones.

Congenital insensitivity to pain is an inherited condition resulting from certain mutations in our DNA. Patients must be born with it and so far it has not yet been found to be transmittable. This means at the moment there is no scientifically recognisable cause to indicate that a zombie really would feel no pain. In reality, or virtual reality as the case may be, when a zombie is having his head hacked off he is in just as much pain as you or I would be. It is more likely that the damage to his vocal processes means he lacks the ability to articulate his horrendous agony, he will obviously moan a lot.

It is more likely that the idea of zombies feeling no pain is a myth perpetrated by video game manufacturers to stop any moral questioning around the daily digital murder dished out to zombies worldwide.

Now that we have ruled put those symptoms currently improbable in an acquired condition like Zombieism we can get a more realistic picture of what a real zombie would actually be like. All of the below symptoms have been found as possible to occur in our bodies via infection or by another acquired method. Let's now examine how they can come about.

- Zombie Walk: Poor balance and coordination resulting in a shuffling, unsteady walk

- Zombie Personality: The loss of speech, memories and other human personality aspects

- Zombie Diet: An unending desire to consume any food source

- Zombie Moan: Communicating only with an incessant moan

Zombieism symptom: Poor balance and coordination resulting in a shuffling unsteady walk

With the exception of a drunken Uncle on the dance floor nothing can match the majestic stumble of the zombie. This famous zombie walk can and does occur in humans frequently due to a range of causes. Think of the human brain, it is the most complex thing currently known of in our universe.8 Although doctors and scientists have been studying the brain for a long time there is much they still don't know about it. It is believed that it functions like a powerful computer, not only controlling the inner workings of your body but processing information constantly coming in via your senses. It seems that certain areas control certain functions. All of the body's voluntary functions are controlled by the brain.

In regard to making any movement the amount of information gathering, planning and decision making going on in our brains is hard to comprehend. Having a sip of tea involves setting in motion a long sequence of muscles at just the right force and speed, we do it without thinking yet it is vastly complex.

The sections of the brain known as the cerebellum and basal ganglia are areas which regulate balance and coordination respectively.

The cerebellum acts like the conductor of an orchestra. It operates to synchronise our movements, to keep them fluid and smooth. Upon receiving information regarding the intended movement the cerebellum sends signals back instructing how it is to be performed.

---

8 Note from Doctor Austin: "After my ex wife the human brain is the most complex thing currently known of in our universe."

Working in conjunction with the basal ganglia the move is then carried out.

The basal ganglia are a cluster of nerve cells responsible for beginning and then regulating our movements. Patients who suffer from Parkinson's disease experience damage to this area. This makes movement shaky and they demonstrate a slowness to start and then execute a series of moves.

Damage to these areas can be caused in many ways. It may be a physical trauma to the head or a disease such as Parkinson's or Kuru but the results are the same; poor coordination, severe tremors, and a wide legged, unsteady lurching walk. These symptoms can also be viewed in most British city centres on the weekend due to alcohols depressive effects on the cerebellum. Any zombie drinking alcohol would become immobile rapidly.

Zombieism symptom: loss of speech, memory and other human personality aspects

The high complexity of the brain makes it awkward to establish exactly which areas are being used during certain actions. If a large amount of areas are affected the end result becomes even less predictable. As we will go on to discover Zombieism creates a wide variety of damage to many areas of the brain. This means occasional unusual symptoms might crop up in those infected with Zombieism. Some may seem more or less human than others. Generally the area's most affected will include:

The frontal lobes: Our frontal lobes are home to our emotional control centres and personality. No other part of the brain is capable of exhibiting such a wide variety of symptoms when injured. Amongst their many aspects these lobes play host to our impulse control. Dr Steven Schlozman describes impulsivity as "if you had a few more seconds, you might not have done that" whereas perfumer Chanel describes impulsivity as "a passionate vibrant fragrance for the modern woman."

Due to its forward location this area of the brain is at highest risk of damage as a man named Phineas Gage found out first hand in the mid 1800s when he survived a metal rod through the face and skull. Many said he was in the wrong place at the wrong time. Although technically he was in the right place at the right time. He was at work on the railroads when an accidental explosion blew an iron spike up and under his cheekbone then out of the top of his head. Miraculously he managed to walk assisted to a cart that took him to hospital. This was after taking delivery of what was essentially a frontal lobotomy. It won't surprise you to learn that this experience changed him. In fact the frontal lobe damage he received vastly changed his personality. Not only did he have trouble finding suitable hats he also became more impulsive and slightly more conductive. Once a hard working, polite and friendly man he had become rude, capricious and untruthful. His colleagues no longer wished to work with him. He had lost his previous social skills. The case of Gage was important for the advancement of the then fledgling science of neuropsychology because it gave good indications about the role of specific areas in brain function.

Those like Mr Gage with frontal lobe damage show some interesting symptoms in the context of Zombieism. A main characteristic is an inability to comprehend their environment. This adds credence to the explanation as to why zombies become accidental cannibals. Another is increased risk taking and non compliance potentially explaining why a zombie appears to play by its own rules. Physically frontal lobe damage can make it harder to make complicated muscle movements. It becomes more difficult to compose facial expressions. This shows how a zombie could be unable to contort his face into a look of pain when you pound its skull.

The sheer range of possible symptoms resulting from varying damage to the frontal lobe makes it a front runner as part of the branches of infection extended by Zombieism.

Amygdala: Attempted murder aside your average zombie doesn't get up to much. Doesn't wile away the afternoons taking windy walks, or read the Sunday Papers by the fire. It appears to act like a creature of instinct.

Coming in at about 1 inch in length and shaped like an almond is a brain structure buried deep within the temporal lobes. It is known as the amygdala. It is from here our instincts and base emotions like fear & arousal come from. Many believe this is the part that gives us our humanity. We humans have animal impulses and instincts but the intellect and reasoning to control them. Our natural instincts are unlearned, inherited behaviours. Two examples are sex and feeding. A fictional zombie's natural instincts hardly seem to fit into the survival of the fittest. Everything about them thus far dooms them to rapid failure. Compared to a standard human a zombie does appear to exist in a more instinctive state. But like a human its primary functions are still feeding and sex. Although to a zombie these are one in the same. These are actually the only two instincts zombies ever show. It does not demonstrate any aversion to danger, another common instinct.

The very rare genetic condition of Urbach-Wiethe disease can lead to degeneration of the amygdala. A 44 year old woman known as SM has this disorder and as a result experiences no fear. This leaves her unable to detect threats in her environment and as a result she does not actively avoid threats as other people would. (Feinstein, 2010) However her other instincts were left intact.

The formation of lesions around this area in feral rhesus monkeys transformed them from wild beasts into tame docile creatures. Other studies with monkeys and the amygdala have shown that the fear response can be inhibited.

A zombie does not run when faced with insurmountable odds yet this is not an act of bravery it is an act of its nature. Somewhere deep down inside it may grasp the concept of fear but it does not know fear as we do.

The amygdala's relationship with Zombieism does not end with instinct and fear. Anger too plays an important role. The amygdala's varying size across species is postiviely collerated with agression. The crocodile has a very large amygdala, so getting angry with a crocodile for being a crocodile is not only dangerous but scientifically silly because it can't help being a crocodile, obviously, because it is a crocodile. Interestingly in human males the amygdala decreases in size by as much as 30% upon castration. This may offer a strategy for calming down large male zombie populations, get a mini guillotine and get to chopping.

Post traumatic stress disorder (PTSD) is believed to involve the amygdala. Patients suffer flashbacks, nightmares, bouts of anger and hyper vigilance amongst other symptoms. It can be brought on as the result of a traumatic events occurring on one's life. There is little doubt that the progression into Zombieism would be an unpleasant and scary one. Especially if the person is aware of what is happening to them and/or has witnessed it in others. During the onset of Zombieism conditions will be rife for PTSD. Furthermore during this person's life with the Zombieism condition they will no doubt find themselves victimised and violently attacked by ill informed members of the public. Daily dices with decapitation and regular run ins with rifle rounds mean that the majority of zombies are likely to have this condition. It may account for further damage going on in the amygdala.

It should now be clear to you that a crocodile is a crocodile and will always be a crocodile and that Zombieism may cause harm to the amygdala resulting in a creature that appears to act purely on instinct.

Zombieism symptom: An unending desire to consume any food source

After a fictional zombie has followed its instincts at a leisurely pace and located a living victim the biting can begin, but once it starts it usually never ends. Zombies consume and consume like Americans in an all you can eat buffet. What is the cause of this?

Whilst discussing cannibalism earlier it was mentioned that damage to the brain can produce a compulsion to eat. Eating habits are controlled from the ventromedial hypothalamus, a subgroup of the hypothalamus. The hypothalamus is a small coned shaped structure. From an evolutionary perspective it is the most ancient part of our brain. It has remained unchanged for a long time. This means it is a brain region present in all birds, reptiles and mammals. As with much brain research we don't know exactly what it does but believe it is roughly involved in; releasing some major hormones, temperature regulation, controlling food and water intake, sexual behaviours and reproduction, control of daily physiological state & behaviour and mediation of emotional responses.

The ventromedial hypothalamus is also known as the ventromedial nucleus and has often been termed as the satiety centre. It is a small and sensitive area. When stimulated its causes us to stop eating. But when damaged or destroyed it can cause us to eat and eat without stopping. Not so far away is the lateral hypothalamus, damage here has the opposite effect making people instead starve to death.

Certain lesions forming upon the ventromedial hypothalamus are known to cause hyperphagia, eating because you never feel full, obesity, the accumulation of excess fat, and increased expression of aggressive behaviour. This terrifying trio can be found in abundance amongst fictional zombies and it would seem real ones as well. Obviously with regard to obesity it is slightly different. A large percentage of the zombie population will likely be obese. This is because the host bodies were obese prior to acquiring the Zombieism condition. A thin zombie won't become obese because its body is not absorbing nutrients in the same way as a regular human body furthermore it wouldn't live long enough to gain much weight.

These lesions aren't the only thing that can cause hyperphagia. Hyperphagic conditions have been related to many sources including; Genetic disorders, psychiatric disorders, sleep disorders, medication and psychotropic compounds like delta-9 tetrahydrocannabinol. You may be familiar with this final substance by its other name THC it is what puts the high in cannabis. In the marijuana world hyperphagia is known by the name 'the munchies'. So to say 'those zombies have the mad munchies' is a completely true statement. To also say that this area of the brain is important for the Zombieism Condition is also true.

As we said it isn't really possible to make a human zombie a cannibal. But with these lesions or some other type of damage they would be aggressively and obsessively working to resolve an imaginary hunger. They would stuff themselves ceaselessly when possible only to purge themselves in an explosive vomiting spree and begin the cycle once again. We also said that a zombie's general lack of awareness, eyesight and manners could make it an accidental cannibal as it seeks out any food source. However messing around near the hypothalamus as the Zombieism condition would arguably give us an additional bonus, the rage. This might provide more reason for a zombie to physically attack a human.

Bilateral lesions that include the ventromedial hypothalamus have long been known to cause expressions of rage in animals. Working alongside the amygdala there are in fact many parts of the brain dealing with emotion. The hypothalamus may be where emotional factors influence physical body functions hence the action of eating or the actions of rage. The name give to chronic condition of irritable mood combined with an increase in aggressive behaviour & tendencies is known as hypothalamic rage.

A zombie already behaves as a creature of instinct thanks to the other damage going on to its brain. But the addition of further damage to the ventromedial hypothalamus could also increase its overall rage and aggression. Not only will it be continually searching for sustenance but it will be doing so in the same manner as Godzilla does his shopping in downtown Tokyo, with extreme prejudice.

What's more the ventromedial hypothalamus could be the centre for heterosexual sex drive in women. Laboratory mice had the effects of oestrogen blocked in that area. They discovered oestrogen in the ventromedial hypothalamus was essential to arousal. Without it the females simply would not have sex. It is possible that lack of or loss of sex drive in women may be a symptom of the Zombieism

ZOMBIE SCIENCE 1Z – THE TEXTBOOK

Condition. This is why some men use the expression 'zombie chick' to describe a girl who has little or no sex drive. Please do not confuse this with the other usage of 'zombie chick' which describes a girl who moves stiffly during love making and moans in a masculine way.

Zombieism symptom: Communicating only with an incessant moan

Finally a zombie wouldn't be a zombie without uttering a sound reminiscent of an elderly gentleman having a late night visit to the bathroom. For Zombiology students the zombie moan is as beautiful as the sound of bird song to an ornithologist or the call of the Raptor to a palaeontologist. Many have theorized, even philosophised over the meaning behind these eerie sounds but our research indicates there might not be any.

The term dysarthria refers to a group of motor speech disorders resulting from a disturbance in neuromuscular control. It can affect the entire manner a person talks by disturbing speech in a variety of ways. The condition leads to differing degrees of weakness, slowing, in coordination and inaccuracy of vocal movements. The end product is speech with abnormal characteristics, which can be unintelligible to varying degrees. (R, 2005)

In a severe enough instance a person with dysarthria might emit nothing but an incomprehensible moaning. Impairments can result from damage to the central or peripheral nervous system.

Many existing brain disorders like Dementia cause this dysarthria condition and even a head injury can bring it on. The infamous zombie moaning may be no more than an unfortunate side effect brought about by the combination of ways in which the Zombieism condition is damaging the rest of the brain.

Summary of module one

We have now reached the end of the first module and as long as you haven't begun to develop the symptoms we have covered thus far you should consider yourself doing well.

Let's recap: we identified all the symptoms associated with the fictional zombies.

They were:

- Dying then coming back to life (referred to as being 'undead')

- The continued decomposition of skin and organs (as per normal human decomposition)

- Poor balance and coordination leading to a shuffling unsteady walk

- The loss of reasoning, memories and other human personality aspects

- An unending desire to consume living flesh with a particular fondness for the brain

- An apparent inability to feel pain or heal

- Communicating only with an incessant moan

Using our deductive reasoning we were able to rule out certain zombie symptoms that are poorly conveyed or completely misrepresented in media and popular culture. We now know a real zombie would not be a dead body that acts as though living, it would not be undead, it would not be a ghoul. It has to be living in some respect, even if somewhat badly. We also know a real zombie could not be continuing to decompose like a human corpse. If it were it would dissolve within weeks but be far more ineffective much sooner.

Rather than becoming a cannibal a zombie would simply become an unfussy eater who occasionally partook of some person and contrary to propaganda a zombie would not be immune to the feeling of decapitation it just sadly lacks the muscular control to express its pain and the vocal ability to explain its agony.

Once these symptoms have been disregarded we are left with a more accurate picture of what a real zombie would actually be like.

Here are the potential symptoms of the true Zombieism Condition:

Trademark Shambling Walk: Its semi-slapstick shuffling walk and disastrously poor balance may be the result of damage to the cerebellum and basal ganglia.

A Loss of Human Personality: A combination of damage to the amygdala, frontal lobe and other areas contribute to the loss of human personality and increase of impulsive and instinctive behaviour.

Sudden Change in Eating Habits: Certain lesions on the ventromedial hypothalamus can result in hyperphagia, eating because you never feel full and hypothalamic rage, an increase of aggressive behaviour. This working alongside the instinctive and impulsive behaviour this explains partially why they are violently driven to eat but as was covered in our cannibalism section there is no evidence a zombie would be human/brains exclusive.

Talking Only By Moan: A side effect of the combination of brain damage caused by all of the above can allow dysarthria to take hold reducing their speech to an unintelligible moan.

Now that the symptoms have been outlined and identified it is easy to spot a common link between them, something that brings all of them together. Take a few moments to consider what it might be. The answer won't surprise you, it's really quite apt. The location in the body that looks to be the source of all these Zombieism symptoms is the brain.

We can now attempt to use this knowledge to predict the initial source of a real Zombieism outbreak and how it could bring about the above combination of symptoms.

MODULE TWO – ZOMBIEISM CAUSES (M2)

Overview of potential causes of Zombieism and which are viable (Fact & Fiction)

Now we've recognized that realistic Zombieism symptoms can all stem from damage occurring in the brain we need to ascertain what may be affecting the brain and how this disease could arrive there in the first place.

In fictional zombie scenarios the cause of Zombieism can vary. Take a few seconds to think of the cause of an outbreak in a film. Not the biting part, the part before that, the initial cause of Zombieism.

An array of answers should be flooding forth from your mind at this point. Perhaps you recall the original zombie movie, Night of the Living Dead. Here it was possibly the fault of a radioactive space probe exploding in the Earth's atmosphere. Across the extensive Resident Evil brand of games & movies exposure to a man made T-Virus is the cause of Zombieism. Then in 2008 classic Zombie Strippers the cause is of course, well, the cause escapes us right now but all students should see the film anyway.

In section two we will briefly examine many of the possible Zombieism causes. As you will discover this list is numerable and highly changeable. Just because a Zombieism cause is found to be unlikely at the time this textbook is printed does not mean future developments might come along to turn it on its head. This is the excitement of studying Zombiology, what may at first seem to be a dead end suddenly becomes more alive than you realise. Whilst it is important to sometimes focus ones avenue of thought in developing a theory one must also keep an open mind. Only a fool would disregard a potential theory, you may find it comes back later to bite you in the bum, literally.

We will now look over some Zombieism sources before going on examine the Zombie Institute for Theoretical Studies primary Zombieism theory, Prion Zombieism.

Bacteria

Bacteria are living things that have just one cell. Like viruses they are able to multiply with incredible speed. A population in ideal conditions may double in size every 20 minutes. They are resilient and adept at surviving in extreme conditions for more than three decades. In 1998 an American Microbiologist worked out that the number of Bacteria on the planet Earth was five million, trillion, trillion. Unfortunately at this point his abacus, coincidently the world's largest, ran out of little beads to slide across. There are many different types and categorisations for bacteria. If we take an extremely broad view for a moment we can allocate them into two categories, Bacteria and Cyanobacteria.

Cyanobacteria are extremely good at growing and surviving. They can also be called blue-green algae and have the ability to photosynthesise. Millions of years ago when life on Earth was just beginning these Cyanobacteria may have helped create all our oxygen. There work is less at the forefront of the Bacteria world these days.

Instead we need to focus on that other type of bacteria. Of the other bacteria there are seven main groups. However these all have further sub groups. It does get rather confusing. They are identified in various ways such as by shape and type of cell wall. Identification in the laboratory is basically a microscopic version of the board game guess who.

As you are no doubt aware bacteria like humans can be both good and bad. There are 'friendly' bacteria out there. Millions of them are working on your skin right now. Like the barrage balloons of old London town in WW2 kept planes away these good bacteria are preventing nasty bacteria from landing on our skin.

In your guts 10s of trillions of bacteria are assisting in food digestion, producing vitamins, fending off disease causing bacteria and much more. Yet despite all this amazing work there are as should be expected some bad apples in the bacteria bushel.

Bacteria that are infectious to humans and cause disease are known as Pathogenic Bacteria. A familiar type to you may be Escherichia coli known usually by its nickname E. Coli. As an infectious disease it gets into the blood causing vomiting, stomach cramps and diarrhoea containing blood. More serious problems can develop further down the line. E. Coli may sound like an evil bacteria but the infectious type is just one strain. Other good strains of E. Coli are working in our bodies right now to digest food. This illustrates our love hate relationship with them. Bacteria, can't live with them, can't live without them. No seriously.

A time that was most definitely bad in our bacteria relationship was during the many plagues. The Plague or under its cheerier stage name the Black Death was responsible for a lot of nasty pandemics across the world. The causal agent thought to be behind this untold destruction is Yersinia pestis a bacterium.

The infectious bacterium was passed to human hosts by fleas and rats. Three natural forms of the plague exist Pneumonic, Bubonic and Septicemic. The bubonic form of the plague acts quickly on person, within a matter of days. They will begin to lack energy, get a fever, headaches and chills. By far the most common sign are buboes or swellings on the skin, starting out white, then red before becoming a dark purple or black.

Millions died worldwide particularly in the late 14th century when hundreds of millions of people were claimed by it. At the time this was virtually a third of the European population. The plague only stopped in the UK in the seventeenth century. However cases are still reported to this very day. Treatment has fortunately improved a lot since it was first discovered. For example we now have cars to bring out the dead with more dignity.

Another interesting factor to consider with bacterial agents like the Y.pestis is that they have potential as a biological weapon. During October of 1914 in World War Two Japanese planes dropped rice and wheat containing Y.pestis carrying fleas over Chekiang Province, China. They did the same after another few weeks. The outcome was a local plague that killed 121 people.

Bacterium can certainly have infectious properties and the potential to cause widespread damage. They might also be adaptable enough to serve as biological weapons. Is there a form that might give us the Zombieism condition?

Well as we saw in M1 the condition has to be causing damage to many areas of the brain. Bacterium infections mainly target areas like the respiratory (i.e. Tuberculosis, Whooping Cough, Upper Respiratory Infection) and intestinal systems (i.e. Cholera, Traveller's Diarrhoea, Intestinal Yeast Infection.)

Of course it does get to other areas as well. Bacterial Meningitis is a severe infection of the fluids that surround the brain and run into the spinal cord. Common symptoms are head ache, fever and a stiff neck. When it gets more serious nausea, bruising and coma are all possible. Survivors of this serious condition have reported hearing problems and even mental retardation.

However what we don't see in bacterial infections in the type of symptoms we need for Zombieism. It may be that these infections, as damaging as they are, simply aren't hitting the right spot for us. As we said there are trillions upon trillions of them out there so bacteria may make for a prime Zombieism cause in the future.

Radiation

From the beginning of zombie films this little blighter became the scapegoat for many an undead outbreak. Radiation is energy that travels across space in the form of waves or high speed particles. A high enough blast of this energy can ionize atoms. Because atoms are parts of human cells this ionization can seriously damage them resulting in conditions like cancer.

The vast majority of background radiation we encounter occurs all around us naturally. Cosmic rays reach us from space, gas is emitted from rocks on the ground, in fact nearly everywhere you can think of. People have endeavoured to add to the amount of background radiation over the years by building nuclear power stations, weapons and using x-rays. Still man-made sources only account for 15% of all the radiation we are exposed to and this is mainly through x-rays during medical treatment and airport security checks. Thousands of staff work in these environments daily and despite Heathrow's Terminal Five looking like a scene from Night of the Living Dead airport staff don't appear to be turning into radioactive zombies, yet.

In situations where much stronger doses of radiation were applied we can begin to get a measure of its effects. America used atomic bombs against the Japanese in World War Two. Modified bomber planes dropped two bombs that were the equivalent of 20,000 tonnes of TNT on the cities of Hiroshima and Nagasaki.

The destruction they caused is difficult to imagine. Almost 2,000 feet above the city of Hiroshima the bomb exploded. Out came a mushroom cloud 40,000 feet tall. Bubbling purple gas turned red and burned intensely. What was once a thriving city was now a burning crater, thousands were killed instantly and thousands more died slowly of radiation sickness. The main causes of death during a nuclear blast are thermal burns brought about as a result of infrared radiation and fatal injuries caused by falling structures. Thermal burns are usually only visible on the side of the patient exposed to the blast itself. Patients might also exhibit Beta & Gamma burns after a radioactive attack.

Those who survive the initial blast may be unfortunate enough to go on to develop radiation sickness. Symptoms include: nausea, vomiting, hair loss, organ failure, weakness and death. The exact selection of symptoms you develop and damage you receive is heavily dependent on what level of radiation dose you have been subjected to. You may remember the case of former Russian spy Alexander Litvinenko. He was supposedly poisoned with the radioactive substance polonium 210. The result was death by acute radiation poisoning.

Radiation definitely paints an apocalyptic picture when we look at its destructive capabilities. The bombings of World War Two did create burning wastelands where disfigured people stumbled around confused and uttering painful sounds. Yet this is not quite the scenario we seek in terms of a Zombieism outbreak. Aside from the horror movie style skin burns, radiation exposure and radiation sickness don't provide the symptoms we are searching for.

This is not however the only way radiation can mess with us humans. What occurred with the Simpsons three eyed fish can also happen to us, we can become mutated. Mutation doesn't occur in quite the same way in reality as it does in the movies. A dead human body rapidly turns into a fictional zombie mere moments after being dosed. Yet in real life mutations the effects are slower and much less outwardly noticeable. They are usually heredity meaning they are passed on to the children of those affected.

The ill effects of mutations that have come about in the time since the nuclear bombings of the 20th century are bared more by the descendants of those involved. It is when these mutations alter an important protein in the body that medical effects can begin to arise. The primary condition found in abundance after nuclear accidents or attacks is cancer. This field of research is under considerable debate at the time of writing and more about the long term effects of radiation exposure is bound to develop over the next few years.

Again radiation has failed to provide the required symptoms needed for Zombieism. This may be ultimately because its effects are often uncoordinated. All areas of the body can be hit in different ways by radiation whereas we are seeking something with the precision skills to mainly go for the brain. Despite being deeply seated in the myth and legend of zombies radiation is an extremely improbable source for the Zombieism condition.

Nanobots

Nanorobtics is the science of building machines at the microscopic level. Nanobots are controllable machines existing on a Nano (10^{-9}) meter scale. You need an immensely powerful microscope such as a scanning tunnelling microscope (STM) to see them. This area has many potential scientific applications and would make its biggest splash in the world of medicine where it is referred to as Nanobiotechnology. This fledging field is examining ways in which to design, manufacture, programme and control these wee robots. It is hoped they might be able to diagnose and even cure medical conditions. Attacking as back up for the immune system they could help combat incoming biological predators.

The phrase nanotechnology ascribes to the concept of building things from the bottom up with atomic precision. The first mention of it can be found in 1959 with Richard Freynman a Nobel Prize Winner in Physics. An obvious difficulty in constructing these mini-machines is the fact that they are so small.

In theory small construction centres called personal nanofactories (PNs) will churn out large amounts of high quality mini-machines. To master this is to virtually control matter at the nanoscopic scale. A major advance for the human race. Students may be familiar with the term 'grey goo'. This encompasses a public fear that nanobots might escape and swarm together forming a grey sludge that destroys the Earth. Scientists haven't really backed this theory up. They point out you wouldn't actually make a self-replicating nanobot. That is just asking for trouble.

Nanotechnology is used widely today in the commercial sector. Nanomaterials can be found in sports equipment and in chemicals developed for the manufacturing industry. Nanoparticles help make things lightweight and strong. The science is also becoming more and more integrated into designing and creating even smaller, faster computer chips.

At recently as 2008 a chemical nanobrain was used to operate eight microscopic machines at once. Scientists tipped that in future you could send these mini medics in to carry out surgery on a tumour. However they will need something to guide them and this is where the nanobrain steps in. Molecular devices were attached to this brain and operated successfully. One device was the world's smallest elevator system. Take note that this nanobrain was the first device of its kind ever to be constructed.

Then in 2010 Nanoparticles were used to successfully treat skin cancer in human patients. This Nobel Prize winning technique is called RNA interference (RNAi). It essentially turns off the harmful cancer genes in human melanomas. Small particles warriors are built and introduced into the blood stream. They sneak past the immune system and track down the tumour sites. Once there they fall apart release instructions to turn off the bad genes. The contents are then carried away in the patients pee. More research is needed before this technique will be available as a treatment option. Doctors using the Nanoparticles weren't always successful in guiding them to where they needed to go.

In science fiction nanobots are a familiar device. Famously in the Michael Crichton book Prey swarms of nanobots escape from a secret test facility. They become predatory and start to evolve.

Eventually they gain the ability to work together to form into human structures.

What you may be beginning to gather about Nanotechnology is that it is still largely theoretical. This means we don't have to worry about it right now. But on the day it does become a viable option the Zombie Institute of Theoretical Studies will be extremely vigilant as to its possible usage for zombie shenanigans.

Magic

Science fiction writer, inventor and futurist Arthur C Clarke wrote as his third law of prediction, "any sufficiently advanced technology is indistinguishable from magic".

In Sam Raimi's cult trilogy the Evil Dead it was reading aloud from the passages of a magical book known as the Necronomicon Ex-Mortis or book of the dead that summoned forth undead forces and allowed them to posses the living. Many believed that magic was behind Haitian zombies (see Chemicals section for more details). Stephen King's classic novel Pet Sematary finds an ancient Native American burial ground as the spot to bury something you want to become zombified. Animals and people rise from the dead in these spots and return home. But they're not the same, oh no, they're different, smelly, aggressive, and in the case of children, talkative psychopaths.

Often is it easy to turn and blame magic when Zombieism is concerned. People want answers quickly and aren't always concerned where the finger is pointing. But as Arthur said magic is simply something we don't yet understand. When magic looks to be the cause of Zombieism the chances are it falls into one of the other sources listed in this section. Otherwise it may fall into a new section we have still to discover.

With the exception of British magician Paul Daniels and his wife Debbie McGee, whose work has been known to cause zombification in audiences, magic is not a plausible cause for Zombieism.

Aliens

The debate on extra terrestrial life is so expansive it could be a course of its own. Actually Alienology 1ET is a fantastic course we would highly recommend. As we will go on to see the right kind of virus might turn us into zombies. There is a possibility that such a virus might be designed to bring about such a change. Why would someone want to turn us into zombies? Well maybe they aren't. Let's get some context.

Our Universe is over 14 billion years old, its observable diameter more than 93 billion light years. It contains more than 100 billion galaxies including our Milky Way, millions upon millions of stars and potentially thousands of life supporting planets. However, possible habitable planets orbiting around stars are very far apart, too far away to consider travelling to in a conventional manner like a chemically fuelled rocket ship. Our nearest star system is Alpha Centauri. A NASA space shuttle would take about 165,000 years to get there, providing birds or ice or the Russians didn't stop you ever getting off the launch pad.

Ultimately the time, risks and energy involved in travelling between stars and planets is, for the foreseeable future at least, highly impractical on any species terms. The science we need to solve these problems may be biological; a possible solution could be to send our essence in a frozen dehydrated form.

Picture this hypothetical situation. Millions of years ago in a planetary system a great distance from us there was an alien race. Like us they lived on a nice green and blue planet similar to the Earth, orbiting a star like our Sun. Our Sun works by fusing hydrogen into helium. When there is no more hydrogen available in the core for fusion the Sun will turn into a Red Giant star. As a Red Giant the Sun will be larger, much larger. So much so that it will swallow the Earth. The Earth will disappear.

When this was about to happen this alien race knew they needed to escape their planet. But as we have learned so far, travelling between stars is difficult, and even they didn't have the technology to blast off and settle on a new planet.

Instead they decided to use DNA packaged in viruses. For the purpose of spreading through the Universe viruses and DNA have several important properties; they have the ability to contain and pass on information, tolerate dehydration and the cold of space, and viruses would be capable of infecting a host when it arrives at its destination.

For their plan, there are some assumptions the Aliens will have had to have made about the other intelligent life they were hoping to usurp. They will have thought that it was reasonable to assume that in our Universe most life will be carbon based. Following the carbon assumption, protein and water are also likely. Life will have to be based upon cells - there has to be a base building block. Finally, due to the limitations of viruses they will have decided to attack the one organ they thought they were mostly likely to find, the brain.

A brain is a good target because they could not have relied on any other physical characteristic of the host to be compatible. Everything else will have been unknowable.

So how do the Aliens infect us with the virus that turns us all into them? Maybe they found a way to get their virus into a comet. We scientists like to call comets 'dirty snowballs' because they are basically a mixture of ice and dust. Comets can be anything from hundreds to thousands of kilometres wide. Comets are mainly water, probably not mineral water though. The light signatures in the tail of comets, streaming out as they get closer to the sun, show clear signs of organic compounds. Although difficult with current technology it may one day be possible to launch organic material into space.

A while ago some scientists theorised, but have now largely discredited, that seasonal flu was caused by a virus being washed down from the upper atmosphere by winter weather, this virus, like our organic matter jettisoned into space by aliens, could be carried within a comet.

Comet dust particles constantly rain from the skies, around a hundred thousand billion particles per year, and some of these will fall on people. Although the chances of you becoming infected with a virus are minute it is a possibility.

Now we understand how aliens might be trying to transform us into versions of themselves with virus filled comets, but what exactly does that have to do with zombies?

Viruses are small and are not big enough to fit all the DNA of an alien. The solution is to only send part of the DNA. Not only will this allow the DNA to fit into the virus but it is not necessary to send it all anyway. It is better to utilise much of the hosts DNA which produces characteristics specifically suited to the environment of the host. Thus only the essence of the Alien needs to be sent.

It is highly likely that we human beings wouldn't be compatible with this Alien DNA so the result of the virus rather than turning us into Aliens would be to turn us into something new, something we could call a zombie.

In the future it may be possible for human beings or an Alien race to send parts of their DNA into space inside a virus. The hope being it infects another carbon based life form making it slowly transform into a version of them. If the host is not compatible with the virus unusual things could happen to them, potentially turning them into a zombie.

As hugely plausible as this theory clearly sounds so far it does lack some major factors for becoming a cause of Zombieism. First of all, the Aliens would have to be part zombie in order to provide us with zombie like symptoms. Secondly and rather importantly we have not made contact with or found evidence of extraterrestrial intelligence as yet. The Zombie Institute for Theoretical Studies has a dedicated department headed up by astrophysicist Doctor Douglas Macdonald. They are constantly monitoring the skies using SETI (search for extraterrestrial intelligence) technology. It will be a relief to know they have not seen or heard anything at time of publication.

Overall the threat of Zombieism by extraterrestrials is very low. Only humans located in the remote south of the USA have anything to fear from aliens.

Parasitic disease

Parasites are living organisms that use other organisms as hosts. They can be single celled or large enough to see with the naked eye. Countless horrible diseases are caused as a result of them. Sleeping sickness, Malaria and Kala-azar are just some of the side effects of parasitic creatures making homes in our bodies. An array of varying tricks keeps the parasites away from our immune systems. They can hide or make use of multiple disguises to evade detection.

There are three main classifications used for parasites that can cause disease in humans.

Protozoa – single celled microscopic organisms that can live free or within a host they can be both harmless and infectious to humans. Protozoa parasites can be passed in the bite of a fly allowing it to infest the blood.

Helminths – multi cellular organisms like flukes and tapeworms, generally visible with the eye in the adult stages of development. Around three billion people are affected by these worldwide every year.

Ectoparasites – organisms that attach themselves to the outer surface of a host and do not contribute to its survival, examples include lice, ticks, mites and fleas.

For our Zombieism studies we want to know if parasites can enter the brain since this is where the Zombieism transformation is taking place. The brain is a difficult place to get into. It is heavily protected. A defence wall between the blood stream and brain fluid known as the brain blood barrier keeps most toxins and parasites out. It takes a lot of parasites working together or a pre-existing hole in the wall to allow any through. A current theory being bandied about is that some parasites are capable of releasing enzymes that break down sections of the barrier.

At present there are two known human brain parasites. The pork tapeworm Taenia solium and the amoeba Naegleria fowleri. The pork tapeworm is the most dangerous tapeworm known to humans. If its eggs are consumed, normally from poorly cooked pork, tiny larvae hatch in the small intestine. From there they move across the intestinal wall and off into the bloodstream. Here they go to muscle tissue, the liver and in many instances the brain. Once in the nervous system and/or brain the resulting condition is known as neurocysticerosis. It is possible for the larval cysts to live in tissue for many years without showing symptoms because the immune system does not identify them as a threat. This tapeworm is the leading cause of bran seizures and epilepsy in the developing world. Their arrival causes lesions to begin to form on the brain and symptoms start out with headaches, muscle, spasms and dizziness. If nothing is done it worsens to loss of balance, confusion and death. Depending on the severity of lesions on the brain surgery may be necessary to remove them. Otherwise Albendazole, a worm treatment drug, can be quite effective.

The free living amoeba Naegleria fowleri is commonly found in water. Instances are far rarer compared to the pork tape worm and as a result far less is known about it. Gaining access when water enters the nose and working themselves into the nasal mucosa these amoebae then set off for the brain. On arrival they are believed to emit an enzyme to break down tissue allowing them to enter. A feast of tissue and neurons awaits and they speedily gorge themselves causing extreme damage in a short space of time.

Inflammation of the brain often results. This condition is known as Primary Amoebic Meningocephalitis (PAM). Human most often catch it whilst swimming. If attempts are not made to deal with it quickly death occurs in 3 – 10 days. Symptoms include severe headaches, vomiting and neck stiffness. Unlike with the pork tapeworm cysts are not seen in brain tissue.

It is unclear exactly what treatment can help with PAM. Several drugs have been known to alleviate the condition on occasion but cases are usually fatal and as we mentioned earlier rare so detailed research is not available. Parasites are always evolving and adapting to our strategies to fight them off.

An attempted relationship has been perpetrated in the media over the past few years regarding the parasitic fungus Ophiocordyceps unilateralis and Zombieism. O.unilateralis has the ability to exert behavioural change in Camponotus leonardi ants. Once infected with O.unilateralis the ant will climb down from its normal home in tropical trees to a specific section in the jungle canopy. It bites into a leaf with a death grip and dies. A few days later and a rather gruesome spike like hyphae spawns out the back of its head. This allows the fungus to sexually reproduce and spreads it spores that it might live to infect another day.

Whilst its ability to change the ant's behaviour and guide it to such a specific spot is fascinating it does not actually compare all that much with Zombieism. Even if you were to encounter an army of millions of these ants they would not pose any kind of threat to you. You are not the leafy luxury land in which the fungus is seeking to reproduce. The ants will not bite you making you join them in their unholy quest. Indeed the media regularly use the zombie card to attract the attention of readers. Never in a fictional zombie film has a band a ghouls gone to a shopping centre, dropped down dead, and then have large spikes emerge from their heads and billow parasitic zombie fungus everywhere subsequently passing on the disease to the uninfected who repeat this process again and then again and again. Although this could make for a rather exciting new twist on the genre.9 Furthermore this kind of parasitic manipulation has not been seen in humans at present.

Another organism that gets thrown around when zombies are mentioned is Toxoplasma gondii. It needs two hosts to complete its

---

9 All film ideas accidently developed during the writing of this textbook are copyright© Zombie Institute for Theoretical Studies

life cycle. With extreme Tom & Jerry irony it is the cat and mouse that have this unfortunate honour. As with our fungus friend above T. Gondii has the ability to exert behavioural changes in its hosts. In rats and mice it makes them appear less afraid of the scent of cat urine. Normally they would avoid this area, and upon meeting a cat know not to return again. T. Gondii alters this behaviour causing the rodents to remain in dangerous pee spots for long periods of time. Obviously this allows a lot more mice and rats to be eaten by the cats and for T. Gondii to continue its cycle into the second stage. One of the areas it can encyst on the brain is the amygdala, as we found out earlier this controls aspects of our emotions such as fear.

If these standard hosts aren't available other mammals, like us people, can be utilised instead. The name for this condition in humans is Toxoplasmosis and most sufferers are unlikely to ever develop any noticeable symptoms. But if the host has a compromised immune system, particularly in instances of HIV, severe symptoms such as nausea, headache, fever, lack of coordination, bewilderment and seizures can take hold. Some theories and research are beginning to emerge on the possible behavioural changes it can exert on humans. In females an apparent increase in intelligence was found, they became more outgoing, conscientious, sexually loose, and kind; whereas with males the opposite trends were true. Really quite intriguingly bizarre results, they also construed that all people tended to be more prone to feelings of guilt.

Like our supposed zombie ants T. Gondii is something that has somehow fallen into the Zombiology radar. Even in a worst case scenario its effects it don't satisfy our Zombieism criteria. You may wish to investigate those described behavioural symptoms with a female friend though.

Parasites, they are certainly horror movie material. But Zombieism material? Not quite, having only two potential brain parasites and with neither stepping up to the Zombieism mark it's a swing and a miss for parasites at present.

Genetic Mutation

Everybody has their own DNA. DNA or deoxyribonucleic acid is the instructions that tell your cells how to make you who you are. It's packed with all your genetic information including how tall or short you are, what colours you eyes will be, and whether you'll prefer chocolate or jelly sweets. Virtually all cells carry a complete set of that organisms DNA with the exception of sex cells. The female egg and the male sperm each carry a set of chromosomes and fuse together creating a new set.

Human beings are made up of trillions of cells. The instructions in the DNA tell these cells to make different things. For example your DNA gave you that face, but it doesn't explain your hair cut. Our trillions of cells are potential prey to nasty viruses and diseases like the Zombieism condition.

It is possible for our DNA to change or mutate. Hundreds of diseases are caused by DNA mutations and these mutations come in different forms. It may be an insertion, deletion or change in the DNA code. One such disease is neurofibromatosis. It is caused by many different types of mutations of DNA. Depending on which mutations are present the symptoms of neurofibromatosis could mild and hardly noticeable or extremely severe. This might explain the appearance of Joseph Merrick, known as the Elephant Man, but does not explain the appearance of the Stay Puft Marshmallow Man.

Many changes in DNA codes occur naturally all the time. A defect that appears suddenly in one person and is passed on to just their children must have occurred due to a mutation. You may have heard of achondroplastic dwarfism. All those who have it show a mutant phenotype. A phenotype is an observable characteristic or trait. Out of 94,075 children born in one hospital in Copenhagen, eight dwarfs were born to unaffected parents.

Does this mean that there could be a 'zombie' genetic condition lurking within our DNA, waiting for the right conditions to bring it to life?

Yes, yes it does. The Institute hasn't identified a case yet but the human genetic code is so long and complex that the slightest change or error can produce unpredictable results. Parents at some time and place may have had a zombie child. Perhaps fear or superstition caused them to get rid of it so science could never study the condition. And if a zombie child did grow up into a zombie man? Heredity is important to allow genetic research to be carried out. No matter how drunk a person gets they will never sleep with a zombie therefore no zombie children and unfortunately no genetic line to study. If you have slept with a zombie and had its child, again it's unlikely you'd tell anyone, except maybe Jeremy Kyle. A downside to these genetic zombies would be the inability to spread

Zombieism by biting they would have to sexually reproduce. They would be purely sex zombies.

Doctor Austin's Genetic Sex Zombie Theory has been used previously as the basis for teenage sexual health campaigns. Adverts featured a young couple in the throes of passion as a caption reading 'no protection, what's the worst that could happen' rolled across the screen. It cut to the same girl pregnant with Doctor Austin popping up beside her saying 'you could have a zombie child growing inside you.' The advert was banned in 49 countries. Post screening studies did indicate that it not only greatly discouraged kids from having underage sex it actually discouraged them from ever having sex.

So these genetic zombies would be non-biters and for our research a non-starter. Genetic engineering falls into some of the same pitfalls but will be important later on when we look at module three, preventing & curing Zombieism condition.

Chemicals

When it comes to Zombieism inducing chemicals a man named Clairvius Narcisse is the probably the best known example there is. He is thought to be one of the worlds few recorded chemical zombies. Clairvius was a Haitian man declared dead in 1962 and subsequently buried only to then reappear in 1980 in his home village claiming to have been a zombie for 18 years. His death was recorded by two Western doctors. The death certificate and cases notes remained as physical evidence. Clairvius was able to answer questions put to him by members of his family, questions only he could possibly know the answer to. Thorough investigations on all fronts found no evidence to disprove his story.

Clairvius recalled being conscious but paralysed during his death. He remembered seeing the doctor cover his face with a sheet. He then claimed to have been taken from his grave and resurrected by a sorcerer known in Haitian culture as a bokor. The reason for his having been made into a zombie was said to be an act of revenge by his brother over a land dispute. The method by which people were made into zombies in Haiti was believed to be the combined use of magical powers and a specially prepared zombie potion.

Zombies are very common in Haitian stories and folklore as well as being part of the Vodoun religion. They believe that a person who dies as a result of sorcery leaves themselves open to become a zombie. Sorcerers called bokor, are said to have the power to bring a person back as a zombie. To do so they capture a person's ti bon ange, the part of the soul that holds a person's character, personality and will. Without this they take on the mental characteristics of a zombie and the bokor is able to command and control them.

When creating the zombie corps cadavre (zombie of the flesh) the bokor must prepare a substance capable of bringing about the appearance of death in the required person. As Wade Davis described in his book the Passage of Darkness Haitians' feared not zombies but rather the thought of becoming one.

Dr Wade Davis is amongst an array of skills an ethnobotanist. In the 1980s he travelled to Haiti to explore theories and conceptions that existed about zombies. He particularly wanted to find out if Narcisse's experiences as a zombie were connected to or even caused by a drug preparation of varying ingredients. A hope was that by discovering a substance that could slow down the human body without killing it useful pharmaceutical drugs could be developed.

He was successful in gaining access to the right sources and as a result yielded detailed documentation on the preparation process and obtained samples of various preparations as well.

Upon analysis of these preparations he made some interesting discoveries.

Firstly that the ingredients found in concoctions differed widely dependant on region. Secondly that three categories of item were usually used in its creation, human remains, animals and plants that are pharmacologically active and finally elements that irritate the skin and eyes.

Most intriguingly of all he found one ingredient that was common to almost all of the preparations, the deadly marine neurotoxin tetrodotoxin (TTX). Tetrodotoxin can be found in species of puffer fish and a number of other animals. You may have heard of Japan's deliciously deadly delicacy Fugu. This fish dish If not prepared correctly can kill you, if done just right leaving behind just the right amount of TTX you experience an amazing tingling and light-headedness, kind of like tripping.

During the first stage of intoxication patients feel a slight numbness of the lips and tongue. This occurs anywhere from 20 minutes to three hours after ingesting the poison. This leads on to a prickly, burning sensation throughout the body. This effect is called paraesthesia. This may be followed by sensations of weightlessness or floating. Finally stage one wraps up with headache, epigastric pain, nausea, diarrhoea, and/or vomiting. Occasionally, some reeling or difficulty in walking may occur.

Now we reach the second stage. Total paralysis sets in. Most intoxicated victims are no longer able to move at this point and even trying to sit will likely be very difficult. Breathing gets increasingly harder and speech is affected. The patient then usually exhibits dyspnea, cyanosis, and hypotension. Mental impairment, and cardiac arrhythmia may occur. Fitting with the description we heard from Clairvius Narcisse, although completely paralyzed, patients may be conscious and in some cases completely aware of what is going on right up until shortly before death. An exact time of death can be difficult to predict. It normally occurs within 4 to 6 hours of initial exposure but this can vary wildly from as little as 20 minutes to 8 hours.

Although this description certainly does conjure up images of zombies it doesn't go all the way in actually making them. Zombification requires more than just the right amount of this special preparation. In fact during his research Davis found that there could sometimes be so much of the TTX chemical in a preparation that it would have killed the person it was given to. Other preparations had so little TTX they could not have been producing the desired effect. Another element was clearly involved in actually creating a Haitian zombie.

The reason given by Clairvius for being made into a zombie was that he was arguing with his brother over a land dispute. His brother subsequently sought out the services of a bokor to take revenge and resolve the problem. People were only turned into zombies in Haitian culture as an ultimate form of social sanction. The firmly held religious beliefs of these people coupled with the way these beliefs influenced society is what allowed zombies to exist.

A common legend is that these zombies were often put to work as labourers on sugar plantations. Remember that next time your add sugar to your tea, it may have been fingered by a zombie. It's worth noting that like a drunken man trying to put a key in a lock a drugged up zombie makes for a pretty rubbish slave. In the end a large number of recorded Haitian Zombies actually turned out to be missing persons, people suffering from mental disorders or to just be Keith Richards on another bender.

Wade published in his findings in the famous books "The Serpent and the Rainbow" and "The Passage of Darkness". They did much to dispel the myths and rumours surrounding Haitian zombies. The stories of Haitian zombies that historically arrived here in the West are steeped in myth, but inside the legend there is some scientific truth. Zombies have and do exist in Haiti and although they do not exhibit the same symptoms as either fictional or the kind of zombies we are aiming to find they are still the only definitive zombies in our recorded history. So far.

Viruses

Viruses are small infectious agents that can replicate themselves only within living organisms, like us humans. Whatever a virus enters becomes a host, the disease spreads and people demonstrate specific symptoms.

Viruses are basically made up of two parts. They have their genes and these are surrounded by protein. The genes have all the genetic information needed to make new copies of the virus. A good way to picture it is to imagine the genes as instructions to make more viruses and the protein as the envelope these instructions are delivered in. Once the envelope delivers these instructions to a cell they hijack the cells machinery and instruct it to produce more virus copies. After the cell is full of copies it dies releasing these virus copies on to repeat this process on another cell and another and another and so on.

There are many viruses and viral diseases you've probably heard of such as the common cold, influenza or the flu to his friends, and mingingest of all the chicken pox. Indeed many viruses already exist that produce symptoms that could be attributed to a zombie.

One familiar example is Rabies. Rabies is a viral disease that causes inflammation of the brain. It is zoonotic meaning it is transmitted by animals, most commonly by a bite. Rabies is almost invariably fatal if treatment is not administered prior to the onset of severe symptoms.

The rabies virus travels to the brain with an incubation period of usually a few months in humans, depending on the distance the virus must travel through the nervous system to reach the brain. This could explain why in zombie films some people turn quicker than others. If you get bitten in the foot the disease has much further to travel to reach the brain than if you took a bite to the neck or were involved in a gang bite. Gang bite doesn't really sound like a proper term. Group bite? Multiple bitings, yes, better.

Doctor Austin's Zombieism Outbreak Top Tip #432: if you should find yourself being pursued by a gang of zombies, trip up a friend, predators always go for the easiest prey. You can always make new friends but there's no cure for zombieism. If they are an overweight friend even better, as mentioned earlier there will be no overweight people after a zombie outbreak.

Back to Rabies, early-stage symptoms of rabies are headache and fever progressing to acute pain, violent movements, uncontrolled excitement, so perhaps some sporadic dancing and singing of 'way hay I've got rabies', then depression, less dancing now and more solemn mutterings like 'oh no I've got rabies', and hydrophobia, which is the fear of water or the inability to quench your thirst...for blood, well maybe not, but finally, the patient much like a 17 year old in Ibiza may experience periods of mania and lethargy, eventually leading to coma. Worldwide, the vast majority of human rabies cases (approximately 97%) come from dog bites. The other three percent come from Ninja Cats. Keep in the mind that Ninja Cats are nearly always behind you and in the shadows.10

So could there be a zombie virus out there already waiting like a sad, slightly unstable, talent show hopeful for its chance to shine?

Current ZITS research indicates this is unlikely. Heck if an unknown virus was out there statistically Russell Brand would have caught it by now. Now I think about it he does demonstrate many of the signs...wait, on reflection the magnificent Katy Perry would never marry a zombie.

However viruses mutate a lot quicker than other organisms. In many cases their genes are a lot less stable so more errors can occur in their DNA code creating new and different viruses. So whilst we don't presently get the range of symptoms we need for Zombieism from a virus their very nature might allow one to eventually arise.

Primary Zombieism Cause: New strain of prion disease

This leaves us with one remaining possibility, that Zombieism could be caused a rogue prion. It is this theory that is currently at the forefront of the research at the Zombie Institute for Theoretical Studies and has been the focus of Doctor Austin's scientific career for nearly his entire life.

---

10 This is a joke. Ninja cats have very good hygiene and health standards. They rarely have rabies.

"Like a rotten apple, once inside the brain, the mutant form of prion protein turns the native protein into more copies of the deviant, infectious form." – (Source: 11th April 1996. Nature)

A prion is a rouge form of protein. The normal form of the prion protein is produced naturally in all mammals and is harmless. However altered forms can become infectious agents. Exactly what functions the normal prions have in the body are still unknown to science.

Normal prion proteins, when folded up, have a very specific 3D shape relevant to their function. Rogue prions mess with this shape causing them to fold up incorrectly. All these messed up shapes clump together until the immune system notices and works to destroy them. There destructions leave behind tiny holes in the brain giving it an almost Swiss cheese like spongy appearance.

Prions diseases come in several strains that cause differing symptoms but have many in common including loss of motor coordination, dementia and death, and the aforementioned brain full of holes. Prion diseases in humans are presently few in number but multiple strains of each exist and new varieties are being found regularly. Examples are; Creutzfeldt Jakob Disease (CJD), Fatal Familial Insomnia (FFI), Kuru and Gerstmann -Sträussler-Scheinker disease (GSS).

There are different prion diseases that affect animals. There is evidence to suggest the human disease vCJD is the result of bovine spongiform encephalopathy (BSE) a disorder affecting cattle. Other animal prion diseases include; scrapie, which affects sheep and chronic wasting disease, which affects deer and elk.

Infectious prions are different from viruses because they are all protein. This is confusing to scientists because the protein in viruses acts as the envelope delivering genetic instructions to make more viruses. This makes a prion appear to be just an envelope with no apparent instructions.

Prions are difficult for anyone to get there head round. They do lots of weird things. A cornerstone of biology is that all living things have hereditary information in the form of DNA or RNA. Even the simplest of viruses contain genetic material. Prions seem to contain none. This appears counter to everything biology teaches us, except of course when it is teaching us about prions. Some sceptics argued that because prions can come in different strains this demonstrates they must have genetic material. However recent research at the Scripps Research Institute has indicated that differing strains might occur due differing ways the prion is folding up. They found that prions can adapt to new environments by self perpetuating structural mutations. So just as mutations in a virus alter its genetic code and create a different version, changes in the way a prion folds can change the way it affects the brain.

The variations occurring in the prions appear to give them a clear evolutionary advantage and make them fall in line with the Darwinian theory of evolution. These large scale mutations have shown that they can adapt to resist drug treatment something only previously seen in viruses and bacteria.

Another strange thing about prion diseases is that you can get them in different ways. They can be acquired as is thought to be the case for vCJD. It is believed to have come about as the result of eating meat infected with BSE. It can be an inherited prion disease (IPD), meaning you are born with a faulty gene from either parent that creates the rogue prions. Iatrogenic is when you get the infection from contaminated medical equipment. Finally, and possibly the most alarming of all, are the sporadic forms. No one knows why the sporadic forms come about and sporadic CJD (sCJD) is the most common type in the UK.

If the Zombieism Condition was caused by a prion disease initially it would be an acquired one. However once it had been in the population for some time inherited occurrences may start to crop up. A more worrying thought is that Zombieism could begin randomly like sCJD. This would take us all by unpleasant surprise. It would also negate the purpose of the Noah Project. (See below)

It is still unknown exactly how the accumulation of prions damages cells. One theory is that prions build up in intracellular vesicles known as lysosomes. In the brain, filled lysosomes could burst, damage cells and spread the infectious prions further. As these diseased cells die they then leave behind holes in the brain. Spongy hole damage can be seen in all prion diseases.

Diagnosing prion diseases in humans is very difficult. It is easier if the condition is genetic because the patients DNA can be tested for the presence of the mutated gene. This comes with its downsides as well. If a parent dies from the condition this leaves their children facing the unpleasant fact that there is a good chance one or more of them might have it as well. The implications for the remainder of those people's lives are extremely serious.

In most cases of prion disease the only definitive diagnostic test is a brain biopsy. A brain biopsy is a procedure whereby a large needle is inserted into the brain to remove a tissue sample. A drill is used initially to create the guide hole for the needle. There is a risk that a brain biopsy can damage the brain and induce seizures. According to the National Prion Clinic the procedure is only performed in a few cases usually where there is a concern that the patient does not have CJD but some other treatable condition. It is obviously easier to perform this procedure if the patient is deceased. However its effectiveness as a method of preventable diagnosis is somewhat reduced by that point to a level of less than zero percent.

Fortunately a recent development has opened up a new route for diagnosis for vCJD. This is particularly pertinent for us because vCJD is an acquired form of prion disease. A procedure called a tonsil biopsy removes tonsil tissue to examine it for signs of the rogue prion. As variant CJD acts differently on the body than the other forms certain markers can be looked for in the tonsil tissue. But a positive result still can't lead a definitive diagnosis. It can only help strongly implicate the presence of a prion disease when used alongside other methods.

An even faster and more accurate method is also being worked on by the National Institute of Health's National Institute of Allergy and Infectious Diseases (NIAID). The technique has the catchy title of real time quaking induced conversion assay, or if you prefer a more complicated acronym RT-QuIC. In a test tube, the assay detects when very small amounts of infectious prions begin convert normal forms into the abnormal version.

They then compare samples from different parts of the body, diluting them and seeing if they still start to convert the normal prions. This lets them work out the relative infectious concentration of each sample.

During their research the scientists used this technique to try to detect prion infections in sheep known to have scrapie and deer suffering from chronic wasting disease.

"In scrapie-infected hamsters, they found surprisingly high levels of prions in nasal fluids, pointing to such fluids as possible sources of contagion in various prion diseases." (Diseases., 2010)

In live animals testing for different strains is also problematic. With scrapie in sheep diagnosis is performed by watching for the clinical symptoms and analysing them based on the subject's age and breed. As with the brain biopsy in humans there are laboratory tests that can be performed on brain tissue. Third eyelid biopsies can also be taken on live animals and analyzed for misfolded protein.

But with non-genetically inherited BSE in cows there is presently no test that can be performed on live cattle. BSE can be diagnosed histopathologically by microscopic examination of brain tissue. In

2010 an interesting study looked into whether a tell tale glowing of the eyes might be able to diagnose BSE. They noticed that the retinas of sheep infected with scrapie gave off a certain glow when subjected to a beam of light from a specialized medical instrument. This may because it is reflecting off of waste materials present in the eyes left by the disease. If this technique can be adapted for BSE in cows we might have a cheap, quick and easy way to make a diagnosis. This method could no doubt then be adapted to locate prion Zombieism in other animals as well.

Until the actual emergence of the prion Zombieism condition it is hard to tell which methods will be best to try and detect it. But there is presently a strong possibility we will not have available those that we need. Developing such procedures is one of the primary aims currently being worked toward by the ZITS.

To learn more about why a rogue prion is the most likely source for Zombieism we will now have an extract from Doctor Austin's bestselling autobiography "Please. Don't Eat My Face."

Doctor Austin on the roots of his Prion Zombieism theory:

"It was about four or five years after the end of WW2 when I was first led to my theory of Prion Zombieism. I was leading an expedition (of one) to Papua New Guinea (PNG) to research the mysterious brain disorder Kuru. Kuru is a rare and fatal brain disorder that was reaching epidemic proportions during the 1950s & 60s amongst the Fore people of PNG. That's Fore pronounced Four-a and not four, there weren't four people who had this. That's not an epidemic. Imagine a zombie movie with only four zombies it'd be like an episode of Big Brother, only more interesting.

This tribe had been cut off from almost any Western influence and gave a fascinating insight into what life might have been like for our ancestors. Nearly the only western thing they'd seen had been fighter planes that crash landed during the wars. My only way of communicating was by way of a local language called Tok Pisin. The only problem was that it was relatively new and everyone spoke different forms. This made conversation difficult. In the many villages I visited I was often the first white man they'd ever seen. I encountered resistance to my attempts at medical examinations. People felt threatened if I wrote down their names or took photographs with my Polaroid camera. It also confused me that they understood I was writing out their names as they didn't seem to read any English. I tried to show them the positive side of my technology. I showed them how use of perspective in a photograph can make it look like I'm holding a mountain in my hands. This was a bad idea. When I held aloft the glossy picture and it developed before their eyes the assembled crowd went crazy shouting and screaming at me. I could roughly make out the words 'Na rausim olgeta samting nogut long mipela' which is part of the Lord's Prayer. My camera was then smashed to pieces and these pieces returned to me. The woman said 'bagarap olgeta' as she handed them to me. This roughly translates as it is completely broken, from the English 'buggered up'.

Over time I eventually adapted to life in PNG. Even my guides were beginning to warm to me. Although I'm sure I overheard one of them say 'Doctor Austin i gat bigpela hevi' to another guide. This means something like Doctor Austin has many big problems. I was also beginning to get to see potential Kuru patients and learn more about the condition. Kuru patients often seemed to have trouble walking and speaking.

As I gathered my data I observed that Kuru seemed to affect members of the same family so initially I thought it could be a genetic disorder passed on via reproduction. It was only after studying their culture & diet in more detail that I hit upon a discovery. The average Fore citizens' diet included human flesh, organs and brains. At that time there were no McDonald's in New Guinea so the only answer had to be cannibalism.

Additional investigation revealed that upon the death of a Fore individual, the maternal kin were in charge of the dismemberment of the corpse. The women would remove the arms and feet, strip the limbs of muscle, remove the brains, and cut open the chest in order to remove internal organs. Sounds disgusting but do bear in mind this was the 50s and 60s so women were still expected to do all the cooking.

Lindenbaum (1979) states that Kuru victims were highly regarded as sources of food, because the layer of fat on victims who died quickly resembled pork. This goes against my findings which found most feast participants felt it was more like chicken. Women also were known to feed morsels such as human brains and various parts of organs to their children and the elderly. This event was particularly savoured by the elderly. Mind you it'd probably been a long time since Gran had gotten some sausage.

Although I didn't realise at the time it was the brain tissue that was highly infectious and transmitted the Kuru by ingestion or contact through open sores or wounds. It was the great researcher Daniel Carleton Gajdusek who first collected brains samples from an 11 year old patient and injected them into the brains of chimpanzees. One of the chimps then went on to develop to the disease. This proved for the first time that the mysterious disease factor was transmissible by infected biomaterial and that it could jump the species barrier. Daniel and a colleague shared in a Noble Prize for their valiant work.

People often ask me why I never worked with these greats in PNG and aided in their discoveries. I had noticed that cases of Kuru were declining as the ruling powers of PNG and Christian missionaries discouraged the practice of cannibalism. I assumed there was a connection. However at that point I was also heavily addicted to opium. One day I took too much and came across a wrecked British fighter plane in the jungle. For the next 26 months I for some reason believed I was an RAF pilot who had been shot down. My guides later told me they tried to bring me back to the village but I kept shouting 'you'll never get me Gerry', throwing rocks (I believed them to be hand grenades) and making the V for victory sign. They don't have that particular meaning for the V sign in PNG. Unfortunately they do have the other one.

By the time I regained my sanity, which was coincidently when I ran out of opium, a lot of progress had been made with Kuru. It was quite handy really as it saved me a lot of bother. We knew that an infectious agent was responsible for Kuru we just had to figure out what it was.

It could be years or even decades before Kuru patients showed any symptoms. This is a common feature of prion diseases and, as I will go on to explain, is why if bitten by a prion zombie you'd take a long time to turn. One Fore bloke took sixty years to show signs. He was the exception rather than the rule but still, it was the longest house call I've ever had.

The name Kuru is derived from the Fore word kuria which means 'to shake'. This is because severe tremors are always seen in the condition. Kuru is also often dubbed 'the laughing sickness' because hysterical laughter is sometimes seen in patients. I never witnessed this particular symptom myself. The first symptoms of Kuru are an unsteady gait, tremors, and slurred speech. Unlike most of the other prion diseases, dementia or a dementia similar condition was either minimal or absent. Mood changes were often present so I suppose this could account for those instances of laughter.

"...to see them, however, regularly progress to neurological degeneration in three to six months (usually three) and to death is another matter and cannot be shrugged off" (Gadjusek DC (1996) Kuru: from the New Guinea field journals 1957-1962. Grand Street, 15:6-33.)

Kuru is currently un-Kuru-able, a small joke there, but seriously there is no cure for this very serious condition. I wanted to find a cure. I really, truly did. But it was cheaper just to say to the Fore people, gonnae11 no eat brains. I remember at the time believing there might be a connection between the symptoms of Kuru and the zombie like symptoms described by Seabrook in his book the Magic Island (he was discussing his experiences in Haiti). But it was the sixties and the opium helped me to believe a lot of things then.

At the time we thought Kuru might have been some kind of slow virus. It wasn't until 1982 that a Dr Stanley Prusiner coined the term prion and we began to understand it more. He'd lost a patient to Creutzfeldt-Jakob disease and this pushed him to learn more about it. He found that that scrapie, a prion disease affecting sheep, and CJD had been shown to be transmissible by injecting extracts of diseased brains into the brains of healthy animals. (Prusiner, Dr Stanley B. 1996. Biology – Fourth Edition) He also got a Nobel Prize for his prion research. I have continued to push for a Zombieism category with the Nobel people but they don't return my calls anymore, hardly very noble.

My next big breakthrough in my Prion Zombieism theory came in 1993 when I was called into to help with the bovine spongiform encephalopathy gig. BSE is a progressive neurological disorder of cattle that results from infection by our unusual transmissible agent the prion.

It has a long incubation period of around 2 – 4 years, much shorter than with Kuru. Sadly by the time you can actually spot any symptoms the cow is within six months of death. The visible signs it starts to show are usually related to behavioural changes. They become nervous and hesitate for longer and longer periods before moving. Something called hyperesthesia can set in causing abnormal increases in sensitivity to stimulation felt by the senses.

---

11 A Scottish word that roughly translates to 'please do not do that' pronounced go-nay-no.

Their walk becomes trembling & unsteady and by the end they usually cannot move at all. They also tend to stare into space for long periods of time although after eight months of observations I determined they do that anyway.

Farmers had told me they could also exhibit aggression toward other cattle as well as humans. I was unsure of this because I had never witnessed the behaviour. That all changed one Sunday afternoon whilst I was rapidly making my way across a field in the rain to get to my car, an Austin Metro that I loved dearly. Without my knowledge the BSE infected cow I had just examined had suddenly jumped up and was running silently in my direction. Before I knew what was happening a muscular hoof grabbed me round the neck and a sharp blade was pressed against my throat. My belongings were taken from my pockets and off she ran frenzied toward my car tipping a fellow cow en route. She smashed the driver's window, totally unnecessary as she had the keys, started the engine and drove off. I never saw my Austin Metro or that cow again. Worst of all my flask of tea and biscuits were in the glove box. The only peace of mind I had was that she would be dead in a few weeks.

BSE began its rise to fame in the United Kingdom around January 1993 with almost 1,000 new cases per week. It was at the peak of its success in 2008 as more than 184,500 cases of BSE had been confirmed in the UK alone in more than 35,000 herds. (Source: Centre for Disease Control and Prevention. 2010)

It is likely that BSE originated as a result of feeding cattle meat-and-bone meal that contained BSE-infected products from a spontaneously occurring case of BSE. This was an important discovery because it is the path the Zombieism condition will probably follow as well. However it was dwarfed by my next discovery which, whilst providing a great advance in Zombieism research, also highlighted the eminent danger from it as well.

The evidence was becoming stronger and stronger for a causal association between the human prion disease variant Creutzfeldt-Jakob disease (vCJD) that was first reported from the United Kingdom in 1996 and the BSE outbreak in cattle. The interval between the most likely period for the initial extended exposure of the population to potentially BSE-contaminated food (1984-1986) and the onset of initial variant CJD cases (1994-1996) is consistent with known incubation periods for the human forms of prion disease, remember the cases of Kuru. (Source: Centre for Disease Control and Prevention. 2010)

Human beings are extremely distant relatives of bovids such as cattle and sheep. Our most recent common ancestor was alive around 70 million years ago. Because of this evolutionary separation, human prions are unlikely to be similar to those of cattle. There seems no compelling reason why humans should contract vCJD from beef.

But this 'evolutionary' family tree is based on general features of the prions -- an overall consensus of similarity. It does not account for the significance of any particular detailed similarity or difference. Therein lies the interest, similarities between human prions and the prion of cattle -- similarities which occur nowhere else in the family tree.

The chance of these two similarities being shared by cattle and humans is extremely remote. These two unlikely similarities are thought to occur in a part of the prion thought to be connected with disease transmission -- presumably, the conversion of normal prions into rogues. (9 May 1996 Nature)

During a visit to the fantastic Eastwood High School in Glasgow I was asked an interesting question by a budding Zombiology student. Why aren't infectious prions broken down in the stomach? If our super strong stomachs are capable of coping with jalapenos why can't they fend off these nasty prions like those which bring about vCJD?

Usually when a foreign agent enters our bodies, through ingestion or otherwise, our immune system recognises it as a threat and attacks it. Prickly prions appear however to be masters of disguise. It may be that they are absorbed through the gut wall where they then pass themselves off to the immune system as something helpful to the body. Once inside they begin to multiply at lymphoid sites and start the journey through the nervous system to the brain.

So the reason that infectious prions aren't broken down in our body is because our immune system doesn't recognise them as a threat.

Rogue prions are resistant wee beasties, bear in mind with acquired cases of Kuru & vCJD infected meat would have been cooked thoroughly first. This also demonstrates why a flame-thrower is probably the worst weapon for a zombie outbreak and why the best is still always science.

The symptoms of vCJD include: changes in gait (walking), hallucinations, lack of coordination (e.g. stumbling, falling), muscle twitching, myoclonic jerks or seizures and rapidly developing delirium or dementia.

It was whilst working with these patients I realised that in many ways these people essentially ended up as zombies. They staggered around like a zombie. The dementia affected their speech making it tonal and incoherent, almost like a zombie moan. On many occasions they appeared lost and confused in their environment. At this stage I had already eliminated those symptoms which were not actually part of Zombieism (as in Module One) so I only needed to find a new or different strain of this prion disease that caused to damage to the ventromedial hypothalamus bringing about the eating changes we need for total Zombieism in a human.

I now knew that different strains of prion disease were possible and because prion diseases cause this spongy damage to areas of the brain different prion diseases could obviously affect different areas. So if a prion disease came along affected the right combination of areas it could produce the Zombieism symptoms necessary for our condition (see M1 for our symptoms).

It was on Friday October 13th 1996 that I had my great realisation. I remember it was a Friday because it was fish and chip Friday at the University, I love fish and chips me. I think it's because I grew up in a fishing village. My whole family were involved in the fishing trade. My Father, Rod, was a fisherman and my Mother was a net. My 16 sisters and I used to swim upstream to get to school, catching our lunch in our teeth. Mind you the chips at the University are rubbish these days I think it's the oil. What was I saying? Oh yes, my great realisation. I realised that day that zombies could become a reality through a Zombieism inducing prion it was only a matter of time.

I rushed straight to the head of the University, well after I finished my fish and chips (they do cost seven British pounds these days), I barged into his office and said, 'the Zombie Institute for Theoretical Studies has done it, we've finally figured out where Zombieism is most likely going to come from'. He jumped out of his seat, slapped his hands on the desk and said, 'I thought we closed that place down six years ago.' I apologised for disturbing him during trousers down time and left. I then told Davey, the Janitor at the Institute, about my discovery and he was delighted. He even did a wee dance with his mop, it was a rare treat.

I then had to figure out where this Zombieism prion was and how it could be transmitted to humans.

A rogue prion depends on being similar enough to the host prion to be able to 'lock in' to its structure and convert it. Transmission works best between animals of the same species. So sheep to sheep human to human. As we already mentioned cannibalism isn't really practiced amongst humans any more. Therefore Zombieism probably won't come from humans. Unless Armin Meiwes is correct in his estimate that they are at least 800 fellow cannibals in Germany alone in which case look out Europe. But as we've heard with BSE and vCJD there is some evidence prions can jump the species barrier, going from cow to human. This makes the most likely source for the Zombieism condition animals.

Unfortunately no reliable, specific test for prion disease in live animals is available. I then knew to find Zombieism I would have to search for an animal that visually exhibited all of the Zombieism symptoms as they do with scrapie and BSE. If I found one I could then attempt to see if it was also infected by a rouge prion. I decided I would have to check one of each gender of animal, so two in total, to be sure.

The Noah Project began 15 years ago and works tirelessly observing all manner of animals 24 hours a day 365 days a year. Except Halloween, it's our Christmas. The Noah Project is the biggest single project ever undertaken at the ZITS. So far we have observed over 5000 animals for signs of Zombieism and you'll be relieved to know that at time of writing we haven't found any yet.

Alongside scientists the Noah Project also allows the public to get involved with Zombieism research. We often have volunteers joining us regularly on the Noah Project. You don't have to be trained or have a science degree to take part. A simple checklist of Zombieism symptoms, a video camera and some warm soup are all you need to start Zombie Animal Spotting. Please note other than a copy of the checklist on our website we don't provide any of these materials. Additionally I am personally working to design an official ZITS certified name badge for volunteers to wear just in case you are questioned whilst out spotting. This stems from a recent incident involving myself whereby the woman next doors husband queried why I was filming in his garden whilst his wife is in the shower. I tried to explain the misunderstanding but the fact my trousers had coincidently fallen to my ankles moments before his arrival didn't help matters. In future the certified badge should help one get out of such tight spots. I'd encourage all Zombiology students to devote their assistance to the project and get friends & family involved as well. The earlier we can spot the dreaded Zombieism condition the more time we will have to study it before it affects a human. Full details of the Noah Project can be found on the Institute's website.

Although it was an arduous, terrifying and often painful journey the rewards were well worth it. I could finally say with near certainty that the Zombieism Condition was going to be caused by a rogue prion. I then had a cup of tea with biscuits (custard creams). "

Summary of module two

We have now reached the end of module two, in summary:
There are lots of factual and fictional sources blamed for Zombieism.
Some of them like bacteria, radiation and brain parasites do not presently provide the kind of symptoms we need for Zombieism.
If Zombieism were a genetic condition it would be limited in its transmission. In order to spread the Zombieism sexual reproduction is needed. This makes it an unlikely source but the recent rise in binge drinking may give it room to manoeuvre at some point.
Other sources like magical spells or cursed books are at present either scientifically indefinable or in other instances where they are implicated it usually turns out to be the fault of another cause.
The futuristic technology of Nanorobtics and the construction of tiny zombie causing Nanobots are something which will need a period of many years to develop to the right level.
A naturally existing Zombieism virus has not yet been discovered. Although viruses have the potential to be a good source of Zombieism and there are viruses that carry a lot of the right symptoms it seems a proper zombie virus will need time to evolve. So will the technology necessary to construct our own designer Zombieism virus.

The latest research from the Zombie Institute for Theoretical Studies and its head, Doctor Austin, indicates the most likely source of Zombieism within our lifetime will be a zombie inducing rogue prion.

This rouge prion would start out affecting an animal. It would then jump the species barrier to infect a human host. The human host would first come in contact with the prion through exposure to an infected food source.

After a long incubation period of years even decades this person would suddenly start showing the signs and symptoms of a zombie. At present there is no exact way to test the person to find out if they definitively have it. Only at a very late stage in the disease's life cycle will the visual symptoms become apparent.

The patients walk would become progressively more unsteady, balance & coordination would falter and a zombie like walk would be the end result.

They would become less and less aware of their surroundings and unable to recall who any one was. Over time they would lose all their standard personality aspects. Their behaviour will become more instinctual.

As a result of losing awareness in their surroundings their attempts to ingest food would become random leading to some accidental cannibalism. The arrival of hyperphagia will cause an almost continual desire to intake any and all available food sources. An increase in aggression could increase cannibalistic tendencies.

Lastly control of speech would falter culminating in the utterance of only unintelligible moans.

After the onset of Zombieism a human would not last much longer than three months (six at a push).

But this depends on multiple factors such as the subject's initial condition, local environment and of course proximity to Bruce Campbell.

Hypothetical Case Study: Patient with Zombieism

Below is a hypothetical case study of a patient with the prion Zombieism Condition as devised by Doctor Austin:

Day 1

Patient, Miss X, was has been referred to the Zombie Institute for Theoretical Studies after failure by other medical personal to discover a cause as to her condition. She first saw her GP complaining of progressive loss of balance & coordination, slurred speech and increase of appetite. Unable to help her she was then referred to a neurologist. After various scans and tests no cause was found. Fortunately the neurologist had some experience with CJD and had Miss X genetically tested. This test showed she did not have an inherited prion condition. It was this neurologist who referred her on to the ZITS. All this took almost two months so what is day one for us is far from day one for Miss X.

Day 18

Miss X was referred to the Institute a few days before Christmas. As we are an academic institution and not a hospital facility we all had two weeks off. We couldn't see her until after the New Year by which point her symptoms had severely worsened. Her walk had become a permanent juddering sliding shuffle. Her eyesight was so poor that at one point she lifted my hand from the table believing it to be a biscuit on a plate. She was not the same friendly Miss X we had met before Christmas. She barely spoke and struggled to recall facts. Her speech had also worsened into a more moan like pattern.

Day 25

Miss X's condition is now extremely ruthless. She cannot have long to live. Her legs no longer function. She drags herself around the room obsessively trying to eat whatever she comes across. She no longer responds to her name, or can engage in any kind of conversation. She seems unaware of her surroundings often staring into the distance. We now believe she has the prion Zombieism condition however only examining the brain itself will be able to tell us if prion disease is present. I have sent several PhD students to search for evidence of all the animal products Miss X has ever eaten. They queried exactly how to do this to which I replied, 'once you find out, all the secrets of the PhD will be yours'. They scarpered off, I love being a lecturer sometimes.

Day 29

Great news, Miss X is dead. Obviously it was bad news for the X family but you have to learn to laugh at these things, just not out loud, at the funeral. We are getting to check out the brain next week once our dissection kit arrives from Ebay.

Day 36

The brain of Miss X shows clear spongy hole damage and I'm certain the tests will reveal the presence of infectious prions. This may be the world's first confirmed zombie and proof that the Zombieism Condition has arrived.

Final note from Doctor Austin: I hope this case study is only hypothetical otherwise it means the Noah Project is a waste of time.

MODULE THREE - PREVENTING & CURING ZOMBIEISM (M3)

How a Zombieism outbreak could begin and spread

Now we've learned the Zombieism symptoms and established their cause is a prion disease we have to look at ways in which we can prevent and cure it. To do this we must discover how our rogue Zombieism prion is getting in and doing its damage. Once we know this we can strategise ways to halt its advance and mop up the destruction it has left behind.

The conventional wisdom on fictional zombies is that they traditionally spread Zombieism by biting people. This act is probably what makes a zombie seem so terrifying. Often people are more fearful of puncture wounds than say a gun shot. This is because we've all felt the pain of skin puncture whether by an accidental cut or stab. We can imagine what this pain would be like amplified. Few of us will ever be struck by a projectile weapon so the feeling remains a physical unknown.

That zombie's bite is not a bad thing. There is another way Zombieism could be spread that makes biting look like a Saturday in the park with a jam sandwich. Some diseases and viruses, like flu, can be spread by coughing and sneezing. When we sneeze up to 40,000 droplets of potentially infectious aerosols are released. Coughing also involves the release of high velocity saliva. Air and fluid transfer is a far more effective way to transmit infection than biting. That is why you should ultimately be more terrified of a sneezing zombie than a biting one.

A bite itself is a wound inflicted by the mouth and in particular the teeth. To study this in better detail we recommend you get a hold of a biologically replicated arm capable of realistically simulating the damage done to human tissue. If you do not have access to this use the nearest equivalent, a Swiss roll. A Swiss roll does have multiple similarities to a human forearm. It has a slightly similar shape and they are both filled with red jam. These similarities increase even further if you pretend your theoretical patient is Swiss.

Now bite into the arm replica/Swiss roll using a movie zombie biting style. You will now have a fair approximation of a zombie bite to the arm. As well as a mouthful of tasty Swiss roll. By examining your cake like wound you will notice that a bite will cause general tissue damage due to tearing and scratching, a risk of serious spurting if a major vessel is damaged, but more importantly allows for infection by bacteria and other pathogens like rabies. Looking through images of any kind of bites doesn't paint a pleasant picture, except when accidental symmetry gives them the appearance of a lovely butterfly.

But how could our Zombieism prion be spread by biting? Well as we saw with Kuru people had to eat infected human brains to catch it. The same theory exists for vCJD. It isn't usually spread by biting. However it can be spread through open wounds. A bite as we saw certainly creates an open wound. A Prion Zombie as we said could become an accidental cannibal and even eat some infected brains. If the brains were infected with the rogue prion they might then stain zombie's teeth. If these zombie's bit you there is a high chance you would most definitely become infected, in anywhere from 10 – 20 years.

Obviously this doesn't sound like an effective way to spread a disease. In actuality biting itself is a poor way to establish an epidemic for a condition like Zombieism. Biting has been used with great success in disease transmission by the mosquito. Using this technique they have spread yellow fever, dengue fever and Chikungunya with devastating results. Yet when it comes to mammals the results are far less promising.

Recall for a moment the viral disease Rabies we mentioned earlier. Can you think of a case involving Rabies whereby a dog had rabies and within 24 hours every dog in the entire country had it, except for a small group of human survivors huddled on the roof of a shopping centre? If you can't worry not it's because such widespread outbreaks never really occur. Although more than 40,000 people die because of Rabies worldwide and a further 10 million people are treated on a yearly basis, it does not reach a serious level.

One important reason for this is because a dog is much larger than a mosquito. When an obviously disease infected animal is angrily racing toward us at high speed, teeth bared, foam dangling we know it's not a cute puppy seeking a cuddle. We know to get out of its way and prepare to administer a firm but fair self defensive kick. The same is true of both fictional and prion zombies. When one of these sad sacks stumbles its way toward you moaning its messed up face off we tend to back away.

Even if a zombie succeeded in biting 30 people at a Twilight convention before anyone realised that shuffling bloke with no arms wasn't Robert Pattinson it is unlikely it would lead to an outbreak. We humans are actually rather experienced at coping with such eventualities.

Take for example SARS, severe acute respiratory syndrome, it began in China. As soon as the World Health Organisation discovered it had spread to North America they clamped down on international flights and reduced travel across borders. Only 43 people in the entire continent died.

There is no big mystery about where a fictional Zombieism outbreak comes from. Look for the people biting the other people and get them. But with Prion Zombieism the genesis of any outbreak would take a different path from that of Hollywood. As we established biting wouldn't give it the kick it needed. The odds of a zombie munching on some rogue prion infected brains and subsequently biting an uninfected person are fairly slim. As we mentioned the Zombieism prion would start out within an animal and jump to humans. The most common way this has occurred previously is by consumption of an infected food source.

In the 2009 zombie comedy Zombieland the cause of their outbreak is blamed on a mutated strain of mad cow disease. Due to the films aim to be entertaining rather than scientifically accurate, and rightly so, they don't go to lengths to explain whether it is a prion disease in their case. It is referred to in the film as a virus but as we Zombiology students knows it is in fact a prion that causes the human form of mad cow disease and not a virus. The frequency of biting to infection in that film indicates clearly shows it could not be. More importantly the initial cause of Zombieism was eating polluted materials, other than not explaining or demonstrating the condition correctly Zombieland is one of the more accurate movies on this subject.

Exactly which food source will produce Zombieism is still unknown. It is hoped that the findings of the Noah Project will be able to answer that question. Upon consumption of this food source many people could become hosts to Zombieism. Once the onset of the condition begins in these hosts biting might begin to feature but its role in spreading the disease will be much less so than portrayed on the silver screen. Its continued spread will be facilitated by random encounters with infected tissues (i.e. contact through open wounds and sores) and by making future generations more susceptible to acquiring the condition.

Now Zombiology student taking into account everything you've learned so far about Zombieism what methods do you think we could use to cope if cases began to get reported? If the worst has happened people are showing onset Zombieism symptoms and Prion Zombies are stumbling around your garden what do you do? What strategies would you use to handle them? Spend several minutes compiling a list before going on with your reading.

Your list should read something along these lines:

- Implement preventative measures

  (E.g. cleaning routines / quarantine / vaccination)

- Cure the Zombieism

- Find a secure location & stay mobile*

- Arm yourself *

(E.g. decapitation, use of firearms / explosives / blunt instrument/ fire)

*The Zombie Institute for Theoretical Studies does not promote or condone these methods no matter how much fun they may actually appear to be.

It's at this point most people, believing the propaganda of movies & popular media would probably opt to start severing heads and smashing skulls left and right. Personal ties go out the window as poor old Granny's skull is popped like a piñata all across the nursing home floor at the first sign of a slight moan. As doctors and scientists we are usually drawn to first attempting to treat a patient before we resort to bashing their face in. Extreme violence is always a tool we have in the back of our minds, sure it is, but sometimes that pesky old Hippocratic Oath holds us back.

The famous motto goes 'prevention is better than cure' and we think it's best to stick by that. The less people that get unwell the more time off we can get. It's a good phrase to keep in mind with regard to Zombieism. So let's now examine the foremost preventative measures scientifically certified to stop a Zombieism outbreak before it starts.

Zombieism Preventative Measure: Brushing the Teeth

By far the number one weapon against zombies has to be the deadly, the dangerous, toothbrush. The toothbrush itself has been around since 3000 BC when it took the form of a frayed twig used as a method to fend off zombies (and decay.) But since the 18th century it has been mass produced in the United Kingdom and now no one has an excuse to be without one.

Its primary use during a zombie outbreak would be for neutering purposes. If all suspected zombies could be made to clean their teeth their bites would lack the Zombieism causing prion. Bites you subsequently receive might be a slight inconvenience but at least you wouldn't turn. Use of a muzzle might be advisable. Regrettably most sufferers of the Zombieism Condition who are at severe onset stage won't be able to brush their own teeth. You will have to do this for them. Also children lack the physical ability to correctly brush their own teeth until around the age of eight, please bear this in mind during your zombie quarantine procedures.

The Zombie Institute for Theoretical Studies Official Teeth Brushing Guide:

What you will need: toothbrush, toothpaste, clean water, teeth

Step 1: Wet the brush end with water then spread it with a light strip of toothpaste. Keep the zombie calm by making pleasurable moans.

Step 2: Grab the zombie by the jaw and squeeze slightly, continue moaning to keep it calm, the zombie should then naturally open its own mouth allowing you to gain access. Feel free to slap that zombie a few times if it misbehaves, it won't really complain and you'll feel better.

Step 3: Start in the back with the upper molars and work in a clockwise direction. Try to aim your bristles toward the gum line. If you happen to have your protractor with you measure it as a 45° angle. Do this for 20 seconds. (The brushing not measuring)

Step 4: Pop that baby in reverse and head away from that gum line. This carries any nasty infectious prions down and out of the zombie mouth.

Step 5: Let yourself naturally glide along to the lower molars giving them a serious go. This is where the majority of brain tissue is likely to gather.

Step 6: Don't forget to get the backs of those teeth. Point the brush bristles up and get in there horizontally to catch anything you've missed.

Step 7: Next comes the biting surface of the teeth. Take these one at a time with a circular motion, every bit counts in the prevention of Zombieism.

Step 8: Just because this is a guide to the teeth doesn't mean you should miss out the cheeks and tongue. These areas should receive a light scrubbing each for about 30 seconds.

Step 9: Finally top it all off by rinsing out the mouth with water or mouthwash. Whilst the most commonly thought of risk during this exercise is being bitten during brushing it is actually this final phase that holds the most danger. To carry out this rinsing Doctor Austin recommends pouring a small amount of water into the subject's mouth, covering it with the hand and then vigorously shaking the head from side to side. He has practised this on his Gran with variable success.

After you have succeeded in neutering all the zombies in your area it is important you then find somewhere secure to keep them. A lockable room is good on a small scale, a prison or mental facility is your best bet on the grander scale. You then need to make sure you give them an uncontaminated food source from then on lest you have to repeat the neutering process again and again.

Carrying a toothbrush at all times and especially during a zombie outbreak is highly recommended.

Zombieism Preventative Measure: Hand Washing

The second best method known by science to prevent an outbreak of Zombieism is washing your hands. All kinds of nasty germs can get onto your dainty digits and then be rubbed into the eyes, nose and mouth as well as spread on to other objects and of course people. Good hand washing is the first line in disease defence and this goes doubly so for zombies. Studies done at the end of 2010 show that hand washing is at its highest ever levels in the United States of America, this may have partially helped avert an epidemic of the H1N1 flu virus. It quite literally is the next best thing to a vaccine.

Doctor Austin on the role of hand washing in Zombieism prevention:

"When I was young there was no such thing as washing your hands. I believe I had read about the practice in a book but never seen it actually performed. It wasn't until the end of the first great war that David Sign, inventor of the sign, created what might have been his greatest work, the please wash your hands sign. It was really quite a marvel. Of course not every house could afford them. Our neighbours, the Maize's, they had one. A whole gang of us piled into their kitchen to see it just as the sun was setting. The way the light caught it made it almost wink at me. It was rectangular and read please wash your hands now. So we did. Everything changed that day.

Now washing your hands is an everyday part of modern life. Yet all too often people neglected to do so and contracting Zombieism is just one of many negatives brought about by failure to wash hands. I recommend that all Zombiology students should learn the best techniques and practise them regularly. My own personal signature move works as follows. Extend the finger you wish to wash. With the alternate soap covered hand curve the fingers to create a cylindrical shape. Place this shape over the extended finger and thrust it back and forth vigorously for about 20 seconds. Repeat for each finger you have. This manoeuvre is named 'plunging'. Please try to plunge yourself at least three times a day. I plunge everywhere I can, whilst lecturing, in my office, even in the university canteen. They have actually requested I stop plunging myself whilst people are eating but I counter with the fact that the dinner ladies don't wear hairs nets on their upper lips a far more heinous crime. I now eat my meals alone in my office, by choice."
Some of you will no doubt be wondering how the miracle invention of hand washing is going to help you when there is a hoard of flesh eating zombies at your door. Quite simply, it won't, not at all. But it will help to prevent them from being there in the first place.

Zombieism Preventative Measure: Math

The last of our top three Anti-Zombieism tools is the ancient and mysterious art of math. By using mathematical models we can attempt to predict the rate at which an infection can spread and how the application of various preventive measures will affect it. Using complex formulas and accounting for as many variables as possible we can produce estimated figures and theorise about how an outbreak of Zombieism might actually behave.
This was first done for zombies in 2009 and its results published by a team of Canadian researchers in the book "Infectious Disease Modelling and Research Progress". They took the traditional

fictional zombie template to use as the basis for their outbreak. Several scenarios were presented including how violent eradication or treatments options would change the outcome.

Three separate classes were created for their model, susceptible (S), zombie (Z) and removed (R). Susceptible covers those who could be bitten by a zombie or die by natural causes. Zombie obviously applies to the zombies and Removed applies to those who have died by zombie attack or some other means. Therefore it is possible for those in the susceptible and removed group to pass into the zombie group.

A lot of factors must be considered in the creation of these types of models. Some people can fight off zombies better than others, zombies can accidently walk off cliffs, new babies will be born, and so on. To describe in complete detail how this process works would be a textbook of its own so we will work more generically with our maths model in this section.

Overall our Canadian colleagues found that a fictional zombie outbreak would have to be dealt with quickly and aggressively to prevent the overrunning of the human population. Even quarantining and curing the zombies didn't make a significant difference in their model.

Unfortunately for our research they were of course basing their assumptions on a fictional zombie. A key difference they described

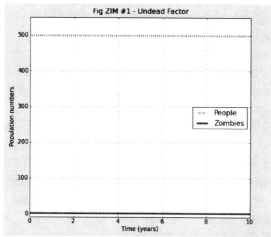

between their type of zombie infection and a comparable existing condition is that zombies come back from the dead. As we have seen this glaring omission will have greatly influenced the final results. If we take into account the fact that the first person to become a fictional zombie would be unable to move, recall the section Dying then coming back to life in module one, we get the following rough picture regarding an outbreak.

In Fig ZIM #1 the dotted line represents our human population, we made this value 500. The solid line represents our zombie population, although it may appear to be at zero it is actually at one. As you can see if one person becomes a fictional zombie, but cannot move in any way, they cannot spread the Zombieism so we will always have just one zombie. Even if we ramp up the amount of zombies that suddenly appear to one million (see Fig ZIM #2),

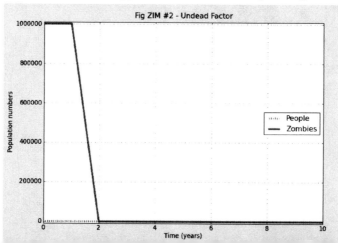

the infection rate still remains zero, the human population is still at 500 and the zombie population is still going remain at one million. You'll notice it mysteriously drops down after two years. Whilst this is most likely caused by an error occurring within the model one could attribute it to the zombie body decomposing over the said length of time. However making such an attribution is simply a good way of trying to cover up instabilities within the model.

This is yet further proof that real zombies cannot be undead, it metaphorically doesn't add up, it technically does.

Here at the Zombie Institute for Theoretical Studies our top mathematician Doctor Douglas MacDonald has attempted to adapt the model for our own uses. He does however make the following disclaimer: "Getting an exact answer using mathematics isn't quite as simple as most people think. At school we get sums such as one plus two equals three. It's all very finite, very definitive. But zombie maths doesn't work that way. One plus two is more likely to equal 3.00000004. Equations themselves are very sensitive, the one used in this zombie model is known as Euler's Method and it can be unpredictable with difficulty in nailing down exact answers. Students are warned that the code in the ZITS model is highly unstable with evidence of chaotic behaviour."

That in mind let's have a look at results gleaned from a ZITS calculated theoretical outbreak. Our three categories work as follows: Unaffected, those who do not have prion Zombieism. Infected, those who have the potential to get the Zombieism condition and finally Zombies, those who are undergoing the three months of full blown prion Zombieism culminating in death.

For our first theoretical scenario, Fig ZIM #3, let's imagine the outbreak begins by the population ingesting a contaminated food source from a fictional burger outlet. We'll name this outlet after our ZITS mathematician Doctor MacDonald. So our fictitious population of around 1000 began eating these infectious burgers from MacDonald's at the start point on our graph.

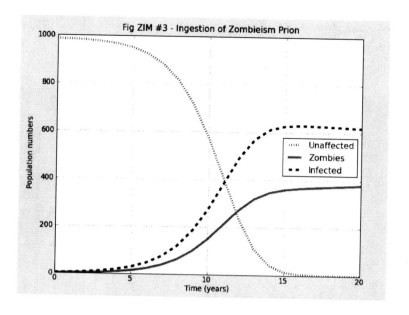

For the purpose of this model we assume that the population don't realise the meat is infectious and continue consuming it at the same rate. As we get into the 10 – 15 year mark we see the most activity with a large chunk succumbing to the disease and becoming Zombies. However a greater amount of humans are carrying the condition but not yet succumbing to it. By the end of the graph we can see that these levels plateau and that whilst every member of the population is affected not all of them are zombies. You would logically assume that someone would eventually realise eating at MacDonald's was causing the Zombieism and burn/close it down.

This theory is often cited in reference to vCJD and would explain why the amount of reported cases per year has decreased steadily since measures were brought in to prevent BSE contaminated meat entering the food chain.

Now assuming our 1000ish population weren't that bright and maintained their level of meat munching. Our steady levels could change once again, as we can see in Fig ZIM #4.

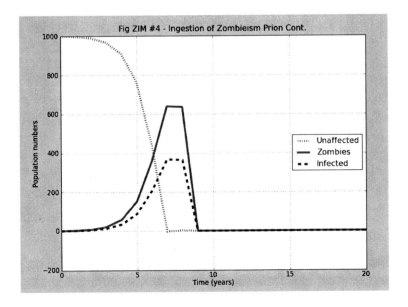

After our outbreak enjoyed some equilibrium the zombie rate shot up turning almost three quarters of our population into zombies. This would make life very difficult for our non-zombie population. It is at this point we would be wishing a cure for Zombieism was available.

We eventually see all our groups fall down to zero. When pressed to explain whether this indicated that ultimately the entire population would be dead our Mathematician replied, "yes, either that or the computer has ran out of RAM."

Now assuming at around seven and a half years into the above outbreak public opinion finally swung in favour of smashing in the zombie's faces would we be able to turn the tide? Obviously killing a zombie isn't the same as curing it. In terms of our model killing a zombie would only remove it from the population without replacing it. So in our scenario even if we killed all the zombies we would only be left with a small amount of humans. Therefore the only weapon that could help maintain the human population and subsequently restore it to its previous levels is sexual intercourse.

If we suddenly add over 900 babies into the population the odds would once again return to our favour. But so do the chances of those babies going on to become zombies. It is a vicious cycle.

These models should make it clear that Zombieism could quickly become a major problem without any method of vaccinating, treating or curing it. Our final model, Fig ZIM #5 leaves us on a more positive note. If we were able to replenish our population through some stupendous shagging, rustle up a cure and do a little decapitating things might just work out for the best.

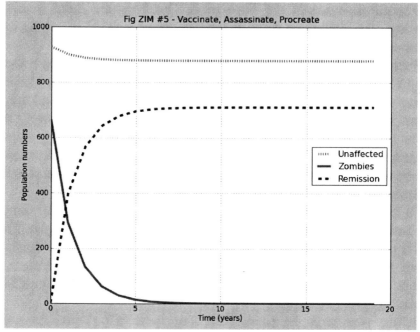

On the plus side if a zombie outbreak did occur on mass these mathematical models should help you convince a member of the opposite sex to sleep with you. Just insist, 'it's the only way we can stop them. We need to make more soldiers.'

If you are interested in learning more about how our mathematical model works please get in touch through the Institute's website, we will be happy to share our calculations.

What would a prion Zombieism outbreak be like?

As there has never been an outbreak of prion Zombieism (that we know of) it is difficult to anticipate exactly what an outbreak would be like. This aspect is often the most ill covered in fictional cases. We usually arrive in the movie to find the majority of people already in the full onset of Zombieism. Where there are survivors they seem to get turned with ease until only the most famous actor is left alive. This probably has more to do with sequels than infection rates.

What we can do is look at how other infectious prion diseases have spread in humans and use this as the basis for comparison. Kuru is a good example to return to. As mentioned it reached epidemic proportions amongst the Fore tribe. The Fore are quite a small group. An Australian Government census recorded a population about 8,000 individuals in the 1950s.

From the year 1957 until 1968 there were believed to be more than 1,100 deaths as a result of Kuru. Women were eight times likelier than men to die of it. Many sources contribute this to women taking part in more cannibalistic feasts than men. This might possibly be due to women outliving men. Children and the elderly were also a more at risk group. Children obviously have more chances to attend feasts over their lifetime and with the elderly the Doctor Austin Sausage Theory applies (see Module Two).

To lose more than an eighth of your population is no small thing. There were virtually no Fore families in PNG unaffected by it. The UK population is above 61 million citizens and rising all the time. If one eighth of the population turned into prion zombies over a ten year period that could see over seven and a half million zombies bumbling around Britain.

Before you begin to panic there are some things you have to remember. The Fore are extremely small in number and their cannibalism was part of their religious culture. They were able to be convinced to cease this behaviour and not all of them took part in it. Prion Zombieism might come about by ingestion of only certain kinds of meat. Not as many people would be as susceptible to encountering it as the Fore people were likely to take part in a feast. As the infectious material is most concentrated in the brain there were cases of people who took part in feasts that did not contract Kuru. This is probably due to which body parts they consumed.

Even with seven million zombies we're still a long way off those zombie apocalypses people are always mentioning. As we will go on to see treatment options might be limited in the case of prion Zombieism. Even if that strain of Zombieism happened to make patients particularly violent and the only most final of final strategy was engaging in public bouts of euthanasia there would be at least 7 people to each zombie. Some would of course be children and elderly people. As zombies they'd be easy prey. But bear in mind that a healthy person beating up someone with prion Zombieism is the equivalent of opening a Paper Mache piñata with crowbar. As they cannot spread the infection by biting there is almost nothing they can do to stop you. Things aren't looking good for the supposed zombie apocalypse. In fact it would look less like 28 days later and more like 28 angry teenagers beating up an infected old woman.

More relevant for us is of course vCJD, the human form of mad cow disease believed to have been brought on by eating infected cattle. This is the most similar case to prion Zombieism seen in humans so far. Up until March 1996 there were only 3 classifications of CJD. As mentioned earlier these were sporadic, familial and iatrogenic. Then variant CJD was identified. It has several differences from the other types. It seems to affect younger age groups, primarily those in their twenties. The average age according to the World Health Organisation is 29. After onset the illness has duration of over a year as opposed to a few months. This could indicate a longer period of Zombieism in the prion Zombieism condition than previously thought.

The first person became ill with vCJD in January of 1994. He was a 19 year old male living within the United Kingdom. Britain has seen the most cases of vCJD in the world and therefore makes it the most likely source of prion Zombieism. This is of course why the ZITS is based in Scotland.

Up until 2002 there had been 129 cases of vCJD reported in UK and 10 in the rest of the world. This is obviously nowhere near as high as the occurrence of Kuru amongst the Fore. It is also clearly not a zombie apocalypse. It is not nearly enough zombie extras for even the most low budget of zombie horror films. The peak of vCJD in the UK occurred in 2008 where there were 28 deaths thought to have been caused by it. Since then it has steadily declined with just 3 deaths reported for 2010. In total there have been 170 deaths from vCJD in the UK as of 2010. If this were Zombieism the odds of you even having heard of it would be extremely slim. The odds of you contracting it even less. None of the extreme violence seen in fiction would be needed.

Some researchers have hinted that there could be a second wave of vCJD in Britain. That some of us could be experience an even greater prior onset period. This comes out of research into our old friend Kuru. Certain key gene shapes have been identified that help fight off Kuru. There are two different forms of the shape. For a change they are easily named, version-m and version-v. Depending on which is present the body will react differently to the Kuru.

Having two Ms (one from each parent) will leave you in the front line for infection whilst getting just one or two Vs gives you a delayed onset. This matches up with vCJD because every patient who has died has been double m. Statistics show around 40% of the population are double m meaning the other 60% may still go on to develop it. Only time will tell with vCJD and so far the real picture is staying down at the bottom end of single figures.

Overall the picture of a prion Zombieism outbreak for certainly the first 10 – 20 years after people start to become zombies is a quiet one. Some might say that reports of a zombie apocalypse have been greatly exaggerated. Or have they?

Thus far our encounters with infectious prion disease have occurred by direct physical contact and ingestion of contaminated materials. Earlier we said you should ultimately be more afraid of a sneezing zombie than a biting one and some recent research for Swiss scientists backs this theory up. Not only that but it may in fact change the zombie forecast completely.

A paper published in January 2011 astonishingly revealed what was previously thought impossible, that prion diseases could travel through the air. More alarmingly this method also allowed them to become more dangerous. A group of scientists at the Swiss Federal Institute for Technology in Zurich developed aerosols containing differing concentrations of brain matter infected with the prion disease Scrapie. These sprays ranged from 0.1% to 20% concentration.

Tests were then carried out on a series of mice whereby they were exposed to the aerosols for varying lengths of time. The results were most unexpected. Virtually all of the mice went on to develop Scrapie. The mice who had been exposed for longer periods, demonstrated shorter incubation periods. As we have covered, prion diseases can take many years to develop. If that was shortened we might find people becoming prion zombies a lot faster than we anticipated.

As the aerosol allowed the prions to sneak in through the nasal passage researchers also believe the infectious prion was actually bypassing the immune system completely instead going straight for the brain via the nasal nerves. Whilst earlier discussing the ingestion of prion infected material we described the journey of the rogue prions and their replication in lymphoid sites before finally heading for the brain. This step seems to be skipped over when it arrives in this manner.

Developing a cure for such a method of infection would be difficult without being able to make use of the resources at hand in the immune system. However it is important to note that people and animals with prion disease do not transmit it infectiously through the air. This research is particularly prudent for scientists working in laboratories on prion samples as well as those working in the cattle industry with infected livestock.

At present no cases of human prion disease can be attributed to airborne infection. There are no unexplained cases that could represent any kind of hidden causal agent such as this. But if our Zombieism prion somehow came in a highly concentrated aerosol we might just get those large numbers and fast turnaround we need for a proper Hollywood style zombie outbreak.

Should prion Zombieism eventually be found within a mammals brain it may be possible to develop an aerosol form such as was used on these test mice. It could then be used as a biological weapon capable of introducing Zombieism quickly and viciously to a population. Obviously no one in their right mind would ever attempt such a thing and we are certainly not encouraging such behaviour. It is best left for fiction.

So with such an uncertain picture, where we might be facing either millions of zombies or just a handful, it's important we are prepared. We should keep in mind that the Zombieism condition is at present a known unknown so things both known and unknown are likely to just possibly somehow happen.   (Johannes Haybaeck, 2011)

Prevention & Cure Of Prion Zombieism: destroying the Prion & repairing the damage

In a prion Zombieism outbreak it appears we have an advantage those in a 90 minute zombie film don't, the luxury of time. By implementing the outlined preventive measures and subsequently preventing future contamination you've bought yourself the time to develop a cure and have to consider two things.

Firstly, how do you remove the infectious rogue prion and secondly how do you repair the damage it has left?

When it comes to treating diseases in general a cure can take on many forms. A cure can be a substance, an operation or even just a change in lifestyle. I'm by no means suggesting Zombieism could be cured by taking up jogging but that doesn't mean you shouldn't do it. It's certainly going to help you stay away from any zombies.

We could try and develop a vaccine to prevent Zombieism from occurring in the first place. A vaccine is a biological preparation that improves immunity to a particular disease. It usually works by injecting a person with an agent that is a weakened form the disease you want to create immunity to. The body works out how to fight it remembers these skills so it can then use them to fight off the real disease in the future. But this would be of little comfort to the already infected.

So how can we begin to treat our prion zombies? And how are prion diseases treated at the moment?

Unfortunately prion diseases are presently always fatal. Treating them is extremely difficult because the brain, like a Thai airport, is a hard place to get drugs into. You'll remember our earlier encounter with the blood brain barrier. Finding drugs will not be easy. The pharmaceutical industry has so far failed to find an effective treatment.

Genetics and genetic engineering could help cure Prion Zombieism.

Genetic engineering is when you make changes to an organisms DNA for any reason. Many people mistakenly assume it's a new thing. Yet from the earliest days of farming humans have attempted to gain control over genetics by doing what is known as selective breeding, breeding plants and animals for desirable qualities. Adolf Hitler of course famously wanted all humans to look like Boris Becker.

In the past this was limited to forms of selective breeding between the same or similar species. As they always say you can send a giraffe and a fish to Paris but they won't make firaffe's. Well now thanks to advances in genetic engineering we can create all kinds of freaky combos, like firaffe's, although no one has yet.

Genetic engineering can be applied to many areas. In industry by engineering genes it is possible to create biological factories that develop insulin for diabetes patients. For over a decade GM crops have been used in agriculture. Crops like corn have been modified to make them more resistant to pests. There are obviously many benefits for researchers. One is that they can genetically modify bacteria to store genes. They become like a memory stick that can hold genetic information. Bacteria can be stored almost indefinitely so it makes gathering and storing genetic information easier. Most relevant for us are the medical applications of genetic engineering. It is this area with the use of gene therapy that could furnish us with a cure.

Gene therapy is all about replacing faulty or mutated versions of DNA codes with the correct copy. It comes in two forms:

Somatic gene therapy which alters the person but is not passed on by reproduction, and germ line gene therapy. Germ line is seen as the controversial or 'Lindsey Lohan' one as any changes made to that person will be passed on forever. Make a mistake that puts our bottoms where our heads should be and you'll have generations of bum faces for decades to come.

Using gene therapy techniques scientists can replace virus genes with human genes they think will cure a disease. One of the most common viruses used for gene therapy is actually HIV because it's so good at infecting, but it's had all the disease causing genes taken out of it. The HIV virus has a protein which makes it really good at getting in to particular cells of the immune system. Similarly the rabies virus has a protein which makes it good at getting in to nervous system cells.

In one of the projects carried out by a researcher now working at the Institute they used a virus that causes the common cold and genetically modified it so that the virus infected heart cells instead. The aim is to develop techniques to treat heart conditions. Theoretically you could use the same method to make a virus which will infect any part of the body you choose. In gene therapy they delete nearly all the virus genes, so they can't replicate anymore and are more or less harmless. They then replace the virus genes with human genes that they think will cure the disease. Theoretically you could use the same science to design a virus to cause a particular disease by altering an existing virus. You could change the protein so that the virus will infect a particular part of the body, as the type of cells that the virus infects will determine the symptoms. And then use either the existing virus genes or add in your own to cause the effect that you want. In future it might be possible to subvert this science to actually produce such a designer Zombieism virus. As we said earlier viruses don't have the same probability for being the cause of Zombieism as a rogue prion does but it is one to watch.

Genetic scientists have artificially constructed lots of viruses. The first was polio in 2002, followed by a recreation of the 1918 influenza strains in 2005. One of the most recent was a bat-borne version of SARS which help scientists learn a lot about it.

Both humans and animals have been discovered to have natural resistance to prion diseases. In 2008 Scientists successfully used gene therapy techniques to treat mice with prion disease. They used specific mutant forms of a mouse gene to reduce the replication of prions in infected cells.

The researchers involved in the study injected this mutant gene into the brains of mice infected with rouge prions. In order to make the study more relevant to human prion diseases, they did this during late stages of the disease, at 80 and 95 days post infection. This increases relevance because, due to the long incubation period of prion diseases, most people are unaware they have contracted them until serious symptoms develop.

They found that, after two injections, treated mice survived 20% longer than non-treated mice. They exhibited substantial improvements in behavioural symptoms, as well as a significant reduction of spongiosis in the brain.

The authors suggest this effect occurred because the mutated gene produces a protein that cannot be converted by the rouge prion. Additionally, the protein delays the work of existing rouge prions by tricking them into landing on these safe prions so they think they are still causing damage.

These results are promising not only because they slow down the toxic prions, but because the effect was demonstrated at such a late stage of disease. Regrettably, the disease was slowed but not cured. Stem Cell research might also hold a future cure to diseases afflicting the brain. This is when damaged cells are replaced with new ones we hope will take their place and heal the damage. To do this special embryonic stem cells are needed. What makes it controversial is that the best sources for these stem cells at present are human embryos commonly those that are left over after IVF treatment. Even though they are usually of lower quality, the best obviously being saved for the future mother, these embryos would otherwise be destroyed. It is these embryos that can then be used for research purposes and to create stem cells.

Stem cells are useful because they continue to divide and repair themselves. We have them in our skin where they constantly repair and replace skin cells. Those stem cells said to have embryonic properties are the most sought after of all. This is because they can become the majority of other tissue in the body. They can become heart cells, skin cells, and importantly for us brain cells. The brain doesn't produce its own stem cells and it can't really repair itself. By parachuting in stem cells to infected areas they can blend in and assume local functions correcting the defect. Gaining command over these cells isn't easy. Scientists have great difficulty in directing these cells to go to the specific place where they are needed. Once they do get there more challenges arise in getting them to form in the type of cell we need and stay that way. There is also a chance the body might reject the stem cells. Solving these complex problems is the basis of stem cell research.

In spite of these benefits there are moral issues surrounding stem cell research. They tend to stem from two areas. The first sensitive issue is that in the development and application of stem cell research human embryos and used and destroyed. Some people think this is akin to murder and the devaluation of human life. The second issue is that through stem cell research we gain the ability to grow fully human clones that could be harvested to provide replacement organs and/or do the things you can't be bothered to.

In early 2008 Robert Lanza MD and his team became the first scientists to create human embryonic stem cells without the destruction of an embryo. They used human skin cells and four proteins. It had only been achieved previously using viruses and genes. Although it was an incredible advance the amount of viable cells turned out was relatively low.

2010 has seen astounding developments as President Obama has lifted restrictions on stem cell research in America. The first trial use of embryonic stem cells in humans has been approved in Atlanta. A biotech company named Geron will attempt to use cells encouraged to develop into nerve cells as a treatment for spinal injuries. Patients who have sustained an injury within 14 days will have the cells injected into their spinal cord. It is hoped damaged tissue will then regenerate. A second trial, designed to help patients with Stargardt's macular dystrophy, an inherited degenerative eye disease that leads to blindness in children, has also been given the go ahead.

As incredible as these discoveries in gene therapy and stem cell research are we are still a long way off from developing a cure for prion diseases. Unfortunately this means there is at present no cure for the Prion Zombieism Condition.

In fictional zombies films humans are usually confronted by these inabilities to medically treat the outbreak. If we were presented with a situation where by prion Zombieism appeared unexpectedly in massive numbers we might be forced down the path of more generic containment methods. Be warned that these methods are by no means as simple as film or computer games have made them out to be. There is a lot more to survival and combat than pressing the correct buttons on a pad in rapid succession.

Last resort Zombieism Strategy: Finding a secure location & stay mobile

Let's look for a start at finding a secure location. In films people usually head for what they think is a safe and easily defendable setting with adequate survival resources. Places like the local shopping centre or pub. It is here we can learn lessons from history. A siege situation was a near everyday occurrence for those living in olden day tall castles. You should be aware that a siege situation only works if there's a chance the invading force will give up. Prion Zombies might only live for three months but the disease can survive and incubate unknowingly for decades.

We humans require many things, mp3 players, the internet, cars, television, mobile phones, bank cards, clothes, Nintendo Wii, money, insurance, good friends, cats, furniture, Simon Cowell, religion, a rule of law, food, water and shelter. The zombie however requires nothing except food, something you both posses and in fact are. By hiding in a stationary location you are merely teasing the zombie. Going 'my milkshake brings all the zombies to the yard', certainly don't do that, zombies instinctively hate Kellis and are angered by all her works. Trapping yourself in a basement or on a roof is perhaps the worst zombie tactic there is.

Instead attempt to use the open landscape to your advantage. As we know a zombie has poor coordination, eyesight and in many cases they would be completely unable to comprehend there environment. Always stay near the banks of rivers or edges of canyons. Upon sight of zombies position yourself on the other side of said river or canyon and simply taunt the zombies.

Traditional & recommended taunts include; 'what's that coming over the hill, is it a...zombie', 'excuse me Mr. Zombie but do you have an alibi, he should respond with a no, and you say, you ain't got no alibi you ugly, you ugly' and finally the Scottish classic 'hoe zombie, smell yer maw12!' That baits them. And like lemmings they will quickly fall or be washed to their deaths.Keep yourself mobile and be ready to start moving the second you notice a zombie coming your way. Memories of your friends being on the receiving end of a gang bite are bound to make you fearful. Don't remember the good times with them. Remember how stupid they were for allowing zombies to surround them in such great numbers. The following combat sections will highlight the important need to keep a good distance between you any zombies you come across.

Be conscious of the transport you choose to use. Prion zombies would likely have some sense of hearing remaining. Whilst they may not have the ability to understand what the loud noise they are hearing is there is still a good chance they, along with any nearby friends, will head toward it. As a result almost all automobiles are ruled out. Preferable methods of transports are; bicycle, unicycle, hot air balloon, segway, foot powered scooter and ideally the stealth bomber. You can anticipate some difficulty in acquiring a few of these. Segway technologies for example, have yet to fully catch on.

---

12 Translated into English: 'pardon me zombie, but would you mind having a sniff of your mother', it is accompanied by a physical gesture in which a hand with index and middle fingers raised is held up and wiggled. The insinuation behind the gesture is that the taunter has physically pleasured the tauntee's mother with his or her fingers leaving oral juices and scenting behind. Confusingly it can be a gesture of both friendship and dislike.

Last resort Zombieism Strategy: Engaging zombies in some form of combat (blunt force / melee weapons, firearms / projectile weapons, fire, explosives)

Blunt Force / Melee Weapons

Assuming that it is impossible for you to avoid a face to face confrontation with the living dead you may be required to engage in physical combat. Sadly it isn't the hilarious slapstick tomfoolery of comedy film Sean of the Dead it is closer to a SAW film in 4D. If you have ever tried to put roller skates on a cat you'll know the difficulty.

Melee weapons are often seen as beneficial because they conserve much needed firearms ammunition. Furthermore zombies are not seen as a physically challenging enemy, unless the outbreak happened at a karate club or Brooklyn.

Weapon selection in such a combat situation is very important. We can't stress this enough. Based on statistics published by OK magazine the most common item that will come to hand during a zombie attack in Britain is the brutal, the vicious, feather duster. To experience this for yourself simply go out and get a feather duster and a cauliflower. A cauliflower is a good stand in for a zombie head because parts of it do look extremely like those of the brain. Give yourself twenty seconds in which to go to town on that zombie beast with the feather duster. Return to the textbook after you have completed this practical experiment.

Welcome back. We hope you found your experience an insightful one. Now as visually entertaining as it may have been it does have some downsides. A large amount of brain matter, tissue and blood has been spread around represented in your case by bits of cauliflower and leaf matter. In prion disease the greatest amount of infectious matter is concentrated in the brain. If you have even the smallest scratch in your skin and accidently get some zombie brain in there you're going to become infected. One of the major ways we think Kuru was transmitted was hand to eye or sore contact.

You may also have noticed that the feather duster itself was a poor weapon for combat. You might have been lucky enough to have a solid handle to wedge it right in there. You may have abandoned it altogether and gone with your fists. If a feather duster cannot adequately destroy a cauliflower what chance has it against a real human skull, the answer, none at all. A big problem with everyday items is there overall inability to crush the human skull. It can take multiple blows to stop all necessary brain activity in the correct areas giving his zombie friends time to join in, well stagger over.

Hand to hand also brings you dangerously close to the zombie. If you haven't been able to brush its teeth yet you may be in for an infectious biting. Some kind of long chopping weapon like a sword is probably the ideal choice. But here at the Institute we prefer to op for non-lethal take downs. Prion zombies would certainly be able to be knocked unconscious. In fact there is no real evidence that fictional zombies can't be knocked unconscious either. It's more that when humans attack them they don't stop until there are bits everywhere.

The human neck contains lots of arteries carrying blood to and from the brain. There are two in particular near the surface that if blocked induce almost instantaneous unconsciousness. Firstly the carotid arteries that stem along the lower jaw bone heading upwards. They carry blood to the brain and away from the heart. Secondly the jugular vein that carries blood from the brain to the heart. This is the easier one to find due to its proximity to the surface. It is about an inch above the collar bone. Both can be found on either side of the neck. Applying a good amount of pressure with one or more fingers in these areas will knock out a prion zombie. Again you will have to get dangerously close to do this and so it is not recommended.

Our findings in the area of melee combat seem to indicate the need for keeping distance between us and the zombie. So let us now look at weapons that are used over distance.

Firearms/Projectile Weapons

Projectile weapons are just the job for taking out a zombie at a distance. Here the United Kingdom the prevalence of guns isn't as high as that of our firearm obsessed Yankee imperialist cousins. Unless of course the Daily Mail newspaper stories are correct in which case every teenager on the street corner has an AK47 hidden inside their tracksuit. Ironically in a zombie outbreak our troubled youths would become one of our greatest assets.

Guns do have the noise disadvantage we covered with cars. This can be averted by use of sound suppressing devices like silencers. Anyone who already owns a silencer is not likely to need that much advice on projectile weapons and may feel free to skip this section.

If we take it that the average student reading this does not have access to a gun (or silencer) other projectile weapons can be found or constructed quite easily.

Doctor Austin discusses homemade projectile weapons:

"By far the easiest projectile weapon to construct in an emergency is a crossbow. I once used my wee nephew's tennis racquet and a piece of flexible wood. I started out by destringing the little blighter, the racquet that is, not my nephew. I then cut a groove in the handle and slid in the piece of flexible wood. I cut another channel down the handle along which the bolts could travel. Then it was simply a case of tying on the string and pulling it back. I used old darts for the bolts. It was certainly most effective for encouraging students to hand in course work on time. I suppose it would work well against a zombie as well."

Images of Doctor Austin's Home-Made Crossbow can be found on the Institute's website.

The crossbow is a good example to use here. As a predecessor to the gunpowder firearm it has many similar traits. Unlike a conventional bow and arrow it does not require much skill to operate. The string is pulled or wound back. As you'll perhaps be aware from far less exciting science of physics this fills the string with potential energy waiting to be released. An arrow or bolt is then placed into the chamber, the device is aimed and the trigger pulled. A modern crossbow is easily capable of sending a bolt clean through the human skull from a quarter of a mile away.

In the event that your only option in dealing with a zombie outbreak is to shoot people in the head the crossbow is really an excellent way to do it. We keep reiterating the zombies are drawn to sound idea. We can't say enough about it. Every time a siren wails near you your head cocks toward it. You know what it is, a fire engine, you've heard it a million times, but you turn your head anyway. What makes you think a zombie wouldn't be at least a little interested in the ricochets of machine gun fire emanating for your location. Crossbows are far quieter, they are the very definition of silent but violent.

Then there's the practising. As newcomers to the study of Zombiology you will likely have had lower than average experience using projectile weapons. We have long advocated the teaching of crossbow skills from preschool age but our calls have fallen on deaf ears. Everyone thinks achieving the headshot is easy. A falsity perpetuated by ever easier computer games. The headshot is difficult and takes time to learn consistently. This means you will need to practice as much as possible when the time comes to take down zombies. If you had a firearm and limited ammunition it would greatly reduce the amount of time you could spend practising. With a crossbow you can practice all day and then retrieve expelled ammo with ease. During subsequent encounters with zombies the same is true. Arrows can be conveniently retrieved from deceased zombie's heads. Although this will be slightly mentally unnerving, but then recycling always is.

Projectile weapons like the crossbow do have drawbacks. Some are very complicated to operate. Larger ones need a lot of strength to set up ready for firing. Guns and rifles get even more intricate. In a stressful situation like a zombie attack you will be scared and shaking. You might quickly find yourself struggling to load a round as the zombies limp closer. The same shaking in terror will also put out your aim as well. As we saw in the previous sections only certain areas of the brain are working so you'd have to know exactly where to hit and be 100% accurate on every single occasion. Any other shot is simply a waste. Prion diseases already create many holes in the brain adding one more might not necessarily do the job. The longer it takes you to put those zombies down the sooner you'll end up back in the trap of hand to hand combat.

Fire

This next tool is often thought to be one of the most effective weapons against zombies. Its acquirement during a video game is usually highly sought after. Yes we're talking about the beautiful and brutal weapon of fire. Most of us have attempted at least once in our lives to construct our own flamethrower. Commonly with a can containing pressurised flammable gas as fuel and a lighter for an ignition source. As comical as memories may be in the event of a zombie outbreak an improvised flame thrower or Molotov cocktail will be about the best you can get. Flamethrowers are no longer in service in the majority of the world's armies. They are commonly very heavy, unreliable and highly dangerous to the operator.

Deodorant cans are usually used as fuel in a homemade device. They frequently contain a combination of alcohol and butane both very flammable gases. The gases are held under pressure and released by pressing a release button. If the venting gas is held behind a lit lighter the result is a roughly one foot long fire stream. If the can itself is heated by fire to temperatures exceeding 170°F the outcome is usually an explosion accompanied by a large fireball. However the risk of the fire stream being sucked back into the can causing immediate explosion is minimal because there is no oxidizing agent in the can to ignite. As we all know oxygen is a crucial part of the fire triangle.

What is of most danger is being burnt by the stream itself or becoming coated in the flammable gas and causing an accidental self ignition. This is a frequent accident among young children who are curious about fire. During instances of constructing spray can flame throwers many children have received severe burns to the skin and more comically set fire to the clothes they were wearing. If you do research on the internet you can observe videos of these weapons in action. Often they are unpredictable and ultimately dangerous. You will then normally see a reaction from the people involved in or near the fire. It is a reaction of panic and fear. In the initial excitement of a zombie attack the thought of fighting them off with a burning torch seems like the greatest strategy on the planet. It is only after everything around us is on fire we recall our human instincts. We are afraid of fire and with good reason. Fire is absolutely terrifying, its hot, its smoky and it kills us. Fire creates epic amounts of fear in humans. A zombie meanwhile remains emotionally unmoved by the experience. Even if we set fire to its head it does not react in anyway. The logic of employing a weapon that induces fear amongst you and your comrades whilst at the same time has no psychological effects on the enemy should be fairly self explanatory.

Any students that have been involved in a spit roast will know another important factor often over looked when using fire as a weapon. It is not an instant kill. Fire needs time in order to do its job. So even if you set a zombie on fire it would take a long time for it to give in to the flames. It could easily be capable of continuing to attack you for two or more hours. Furthermore there is the risk of it then setting you and your friends on fire in the process. Picture this situation. Husband, wife & child are backed against the wall by zombies, all seems lost. Suddenly husband finds spray deodorant and a lighter, he believes all those childhood moments he spent being a pyromaniac will finally come into their own. He burns the zombies, flames lick at their clothes before creeping around and overwhelming their bodies. The man turns to his wife, he is masculine, he is tough, and he has protected his family unit. But then his good ladies look of awe is replaced with one of terror. He turns to find the zombies still advancing not a care in the world. He takes his wife and child by the hand and attempts to escape. But where there was a wall of zombies there is now a wall of red hot burning zombies. That great moment of masculinity is quickly over as the zombies hug the man, his wife and child in a way only they know how. It is doubtful that as his family line ended in burning flames the man still held the belief fire is cool.

Always remember zombie students, like the American Air force fire cannot distinguish targets. Fire is not recommended for use as a weapon during a zombie outbreak at any time. If you do find yourself on fire as a result of a misguided attempt to use it follow these steps:

Don't run around, the fire loves the excitement of it all and gets over stimulated causing it to spread quicker. Instead lie down so the flames find it harder to catch on. Now you must smoother the flames, not with love but with a heavy material like a coat or jacket. You may also wish to roll around slightly too smoother any remaining flames. Finally check to see no one saw you make an utter fool of yourself and/or slap around the person who caused you to get set on fire. This might need to wait until after you have retreated to safe distance.

Fire is also commonly thought of as a useful tool in a zombie outbreak for the safe disposal of bodies. This technique is sometimes applied in infected livestock as in cases of foot and mouth disease. It is only normally done in situations where burials are not possible. This is because of the complications involved in carrying it out and the potential risks from the smoke it gives off. It takes several days to set up the pyre. Cattle are then piled on and coated with a propellant. Burning can last for as long as two days. During which time toxic carcinogenic chemicals are being released in the smoke. The remaining ash is then buried. American farmers have managed to reduce to process from several days to 60 minutes by using napalm.

Applying heat is one of the simplest ways to kill microbes and viruses. High temperatures do their damage by destroying organic molecules such as proteins, carbohydrates, lipid and nucleic acids. These molecules have important roles for the proper functioning of cells and viruses. However prions are a different case. Prion diseases are highly resistant to dry heat, fire, alcohol, boiling, ultra violet & ionizing radiation and most disinfectants. The only method capable of destroying them would destroy their host at the same time. Prions seem nearly indestructible and can endure for very long periods of time.

Overall we advise with fire, as with everything else, that prevention is better than cure. The drawbacks of using fire in any capacity during a zombie outbreak far out way the benefits. It might look and feel cool but the end results are anything but.

Explosives

What about explosives? Playing Grenade bowling with a gang of zombies might sound like fun but what about the scientific reality? Well in a zombie outbreak as with other military kit you will have difficulties getting your hands on any fancy explosives. Inevitably you will have to improvise. Thankfully in part to the rise in global terrorism knowledge about building these improvised explosive devices or IEDs as they are known in the business, is at an all time high.

They've be around a while. The Vietcong used them against US forces. By inserting a grenade and spoon inside soda cans they could make a bomb that exploded when disturbed. They then left these cans in the path of oncoming US soldiers. The age old Western habit of kicking cans across the street came into play and one would inevitably make a play for it, losing a lot more than a goal.

If you've never worked with the Vietcong we can suggest an alternative non lethal bomb to use during a zombie outbreak. Get yourself a can of carbonated (fizzy) juice. Open it just enough to hear the air out but not let the liquid escape. Shake the can vigorously for thirty seconds to a minute. Be careful not to over shake. That may cause premature explosion. Now aim it at your zombie pull back the arm and fire. Run to a safe distance and take cover. The resulting fizz explosion will be immense and cause a massive distraction to confused zombies. If you are lucky some shrapnel damage may occur. The subsequent stickiness and slipperiness of surfaces should slow down any zombie advance. Humming the Benny Hill theme will lighten the mood as zombies hit the deck left and right. It just goes to show that a 24 pack of fizzy cola can quickly go from liquid refreshment to liquid explosive in the event of zombies.

If you're fonder of rocket launchers and rocket propelled grenades then another easy solution awaits. Locate or break off a section of PVC piping. Now track down some home use fireworks, you'll want the rocket ones obviously. Attacking a zombie with a Catherine wheel is simply ridiculous. Insert said rocket into the end of the piping, aim it at a zombie and ignite the fuse. Bam, you have your very own RPG. Actually it is more like a mortar but RPG sounds a lot more exciting. If you want to get a bit more advanced you could construct something more elaborate. Again with strong PVC or metal piping you can create the firing tube. A sparking device, like those found in kitchen can be inserted into the tube to form an ignition system. The back end should be covered with a valve that allows gas to be filled into the tube but not escape out. Some kind of flammable gas such as butane is recommended as the fuel. Now load it from the top like an old school musket. Softer projectiles like potatoes can be used. In more threatening situation a big bag of nails would be crude but efficient. After they have been pushed down the shaft you fill the chamber with gas, click the sparker and there you have it - fire in the hole.

We must reiterate at this point that at no time should any one attempt to actually do this or any other method mentioned in this textbook. There are far too many dangerous variables to contend with and it's a much better option to simply retreat to a safe distance when zombies arrive on your doorstep.

That said, when an explosion goes off a variety of factors come into play. The blast itself emits pressure. Depending on the force of this pressure you may be killed at this point. This might be followed by a fireball. Fireballs cause injury by thermal radiation and engulfment. Persons totally engulfed by fireballs are presumed dead by fire fighters. Finally the fragmentation of the explosive device and objects surrounding it is probably the biggest killer. Explosive devices do their main damage by scattering shrapnel over a large area. This is very good for killing humans. Stray shards and pieces can slice open organs and do all kinds of damage eventually leading to death. There is also the danger an explosion creates in your surroundings. It may ignite secondary explosions and cause debris to fall. Studies have shown a high correlation between the proximity of a person to buildings and other objects during an explosion and serious injuries sustained.

Zombies however can still remain combat ready after an explosion. They are using fewer parts of their bodies than you are. The odds of you destroying the required parts with an explosive device are very low. As with fire it will not give you any kind of psychological advantage over them. If not killed zombies may sustain injuries such as ear drum rupture after an explosion. This could be beneficial in allowing you to start making more noise than before.

If you do have an appropriate use for explosives in a Zombieism containment capacity, such as to destroy a bridge to slow there advance, step by step guides are available online.

Summary module three

That brings us to the end of module three.

In the final piece of the Zombieism puzzle we have seen that a zombie outbreak would be a lot different than its silver screen counterparts. It wouldn't be the nerd Holy Grail of zombie apocalypse they're all saving up their pocket money to buy imitation swords for. Quite the opposite it would likely be something on a much smaller scale.

Ultimately treating prion diseases is extremely difficult and although a lot of progress is being made in researching prion diseases they are presently always fatal.

The areas of gene therapy and stem cell research offer some of the best possibilities for future treatments but a lot of work still has to be done.

This means as yet there is no cure for the prion Zombieism condition.

The best way to stop Prion Zombieism is to never let it start. By observing good hand and teeth hygiene you can significantly reduce your chances of being infected. Prevention is always better than cure.

Plan ahead by using logic and math to anticipate what you might face.

In an emergency situation and only as a last resort in coping with an abnormal and completely unprecedented extreme Zombieism outbreak;

Get to a good open location. Avoid a siege situation and use the landscape to your advantage. Stay portable and be ready to move at a moment's notice. Don't forget to stay silent but violent.

Think carefully about which weapons will work best. As we say in our nursery level education leaflets, 'remember kids always carry a weapon capable of easily bludgeoning the human skull'. A full range of Zombieism Condition educational leaflets are available for download on the Institute's website.

Look at the big picture. Hollywood might have glamorised looting and random violence during zombie outbreaks and yes it's bound to be fun decapitating people you owed money to and don't like at first but you will start to miss the comforts of the modern world. You will need to cure the Zombieism eventually.

CONCLUSIONS & COURSE SUMMARY

Recap of all three modules

This brings us to the end of the textbook for Zombie Science 1Z.
We will now briefly review the conclusions from each of the three modules.

In the first module we examined the symptoms associated with Zombieism. We found that many of the symptoms attributed to the condition by films & popular culture are in fact unlikely. Those which are likely stem from diseases affecting the brain.

The second module then looked at the ways this disease could come about in humans. Briefly looking at the most commonly thought of areas to examine feasibility. The primary method discussed and analysed was a prion disease. Although the Institute believes this to be the main cause at present this doesn't mean another area might not develop into one in the future.

Finally module three looked at methods to prevent and treat a prion Zombieism disease. Unfortunately they are currently incurable, and methods like gene therapy and stem cell research are still developing. We also discovered a zombie outbreak would be very different from those in fiction. But just in case Zombieism does occur at an unexpected level we also revealed some last resort methods to cope with a zombie outbreak including physical combat, projectile weapons and fire.

What conclusions can we draw about the Zombieism Condition?

Zombieism Conclusions with Doctor Austin:

"Congratulations on making it to the end of this textbook. You have taken the important first steps in your journey to ultimate Zombiology knowledge.

The zombie touches everybody's life in some way. Young children dress up as them for Halloween, teenagers terrify themselves with horror movies, adults debate the endless merits of weapons for their coming zombie apocalypse and even the elderly sometimes remember hearing stories of foreign lands where creatures of the undead walked.

The fact that awareness of Zombieism is at its highest level is most certainly a good thing for the fledgling Zombiology student. It makes gaining access to funding easier. But it brings with it a whole host of others problems. Misinformation being the greatest of all. As you have now discovered the reality of a zombie is a far cry from those seen in popular media. These are hard truths for the public who have been brought up believing zombies are humans returned from the dead somehow capable of operating without a working brain. The moment a case comes up their first instinct will be to destroy the patient's brain. To them violence is the automatic and final answer.

Strange diseases have evoked these types of feelings in the past. We all remember leper colonies. But even quarantining zombies and giving them cheery bells to wear is not considered an option by most. Fighting these misconceptions will be a big part of your career in Zombiology.

It is understandable that the real Zombieism Condition would make for very different movies. They might not be as fun or as funny so we can forgive people who want to have more creative licence with the condition for the purposes of pure entertainment. But we can't let this blind us from the real threat zombies actually pose to our health.

The next challenge is the rogue prion itself. Understanding it has been extremely difficult to medical science so far. It is a confusing beast and taming it will be tricky. Unlike zombies the public are much less aware of prion disease. This is partially because it does not have as high a mortality rate as other more well known conditions like AIDs and HIV and hasn't been made into a film starring Tom Hanks.

Once we do understand it our ability to develop treatments will dramatically improve. As genetic engineering advances new applications will become available. The entire way we treat disease will change. All of this will bring us that bit closer to curing Zombieism.

I myself abhor violence. I realise I know how to construct more homemade anti zombie weapons than your average bloke but that doesn't mean I'd ever use one. The use of such strategies should be held back as only a very last resort. Try and go with a non-lethal option. The fizzy juice bomb and knock out techniques are ones to remember.

By studying Zombiology your mind should become open to new ways of thinking. New ways of tackling problems and developing solutions. Science has helped us understand most of the easy stuff, now we need to work hard to answer the difficult questions. It is within this domain that the Zombieism condition lies.

Whatever aspect of Zombieism you end up studying and whatever career path you choose in the industry know that Zombieism like Super AIDs may be one of the 21st century's most dangerous diseases.

And if you're reading this as a non-scientist don't forget that you too can contribute to the great work of the Zombie Institute for Theoretical Studies. Get online, join the Noah Project and discover the many of ways you can get involved in the study of Zombiology.

Good luck and science's speed,

Doctor Austin ZITS BSz MSz DPep"

Details of online exam

In order to obtain a pass for the course Zombie Science 1Z and gain the 20 possible credits all students are required to complete an online examination. Please do not forget to complete this exam. If you do so it will result in an automatic fail for the entire course. The examination, which is multiple choice, can be found by going to www.zombiescience.co.uk and clicking on the online examination tab.

END

Bibliography

(NCJDSU), T. N.-J. (2010). The National Creutzfeldt-Jakob Disease Surveillance Unit (NCJDSU). Retrieved from The National Creutzfeldt-Jakob Disease Surveillance Unit (NCJDSU): http://www.cjd.ed.ac.uk/

al, A. e. (2010). Fluorescence Spectroscopy of the Retina for Diagnosis of Transmissible Spongiform Encephalopathies.

Anderson, P. W. (Director). (2002). Resident Evil [Motion Picture].

Biosecurity, C. f. (2010). Fact Sheet: Yersinia pestis (Plague).

Boll, U. (Director). (2003). House of the Dead [Motion Picture].

Brown, K. a. (2010). The Prion Diseases.

Capcom. (2010). Dead Rising 2 (VG).

Choices, N. (2010). Prion Disease. Retrieved from http://health.independent.co.uk/index.php?option=com_nhs&id=870&Itemid=43

Coghlan, A. (2006). Seat of female libido revealed. New Scientist .

Darabont, F. (Director). (2010). The Walking Dead (TV) [Motion Picture].

David Westaway, J. M.-W. (2010). Rapid End-Point Quantitation of Prion Seeding Activity with Sensitivity Comparable to Bioassays. PLoS Pathogens, 2010; 6 (12): e1001217 DOI: 10.1371/journal.ppat.1001217.

Davis, D. W. (1985). Passage of Darkness/The Ethnobiology of the Haitian Zombie.

Davis, D. W. (1988). The Serpent and the Rainbow.

DC, G. (1996). Kuru: from the New Guinea field journals 1957-1962.

Diseases., T. s. (2010). Early Detection Is Possible for Prion Diseases, Study Suggests. Science Daily .

Dr. Sunil R.Lahane, D. P. BOVINE SPONGIFORM ENCEPHALOPATHY : A PERSPECTIVE.

E. Bar-On, D. W. (2002). Congenital insensitivity to pain: ORTHOPAEDIC MANIFESTATIONS. THE JOURNAL OF BONE AND JOINT SURGERY .

Feinstein, J. S. (2010). The Human Amygdala and the Induction and Experience of Fear. Curr. Biol. DOI: 10.1016/j.cub.2010.11.042.

Fleischer, R. (Director). (2009). Zombieland [Motion Picture].

Fox, D. J. (2010). Practical Guide to Neurologic Diseases of Farm Animals and Horses.

Guttman, B. G. (2006). Genetics. Oneworld Publications.

Institute., T. s. (2010). Prions Mutate and Adapt to Host Environment. Science Daily .

Jaume Balagueró, P. P. (Director). (2007). Rec [Motion Picture].

Johannes Haybaeck, M. H. (2011). Aerosols Transmit Prions to Immunocompetent and Immunodeficient Mice. PLoS Pathogens.

Karine Toupet, V. C.-G.-F.-M. (2008). Effective Gene Therapy in a Mouse Model of Prion Diseases PLoS ONE, 3 (7), 0- DOI: 10.1371/journal.pone.0002773).

King, S. (Writer), & Lambert, M. (Director). (1989). Pet Sematary [Motion Picture].

Klitzman, R. (1998). The Trembling Mountain: A Personal Account of Kuru, Cannibals, and Mad Cow Disease.

Lee, J. (Director). (2008). Zombie Strippers! [Motion Picture].

Mark E. Davis, J. E. (2010). Evidence of RNAi in humans from systemically administered siRNA via targeted nanoparticles. Nature.

Matthew P. Frosch, M. P. (2006). Diagnosis and Classification of Prion Diseases.

McKie, R. (2008, August 3). Warning over second wave of CJD cases. The Observer .

NeuroPrion. (n.d.). NeuroPrion. Retrieved from NeuroPrion: http://www.neuroprion.org/en/home.html

Organisation, W. H. (2010). Variant Creutzfeldt-Jakob disease Fact sheet N°180. Retrieved from WHO: http://www.who.int/mediacentre/factsheets/fs180/en/

Page, Sean T. (2010). The Zombie Handbook UK. Severed Press

Page, Sean T. (2011). War against the Walking Dead. Severed Press

Palmer, R. (2005). An Evaluation of Speech and Language Therapy for Chronic Dysarthria: Comparison of conventional and computer approaches (Thesis).

Philip Munz, I. H. (2009). WHEN ZOMBIES ATTACK!: MATHEMATICAL MODELLING OF AN OUTBREAK OF ZOMBIE INFECTION.

Pirro, M. (Director). (1991). Nudist Colony of the Dead [Motion Picture].

Prevention, C. f. (2010).

Prusiner, D. S. (1996). Biology – Fourth Edition.

PRUSINER, S. B. (1998). Prions.

Raimi, S. (Writer), & Raimi, S. (Director). (1981). Evil Dead [Motion Picture].

Reitman, I. (Director). (1984). Ghostbusters [Motion Picture].

Rhawn Joseph, P. Hypothalamus From: Neuropsychiatry, Neuropsychology, Clinical Neuroscience (Lippincott, Williams & Wilkins).

Romero, G. A. (Director). (1978). Dawn of the Dead [Motion Picture].

Romero, G. A. (Director). (1968). Night of the Living Dead [Motion Picture].

Rosalind, M. B. (1998). Fatal Protein: The Story of CJD, BSE, and Other Prion Diseases. Oxford University Press.

Russell, P. J. (1994). Fundamentals of Genetics. HarperCollins College Publishers.

S, L. (1979). Kuru Sorcery. Mayfield Publishing Company.

S. P. Mahal, S. B.-K. (2010). Transfer of a prion strain to different hosts leads to emergence of strain variants. National Academy of Sciences; DOI: 10.1073/pnas.1013014108.

Schlozman, D. S. (2009). A Harvard Psychiatrist Explains Zombie Neurobiology.

Society, A. C. (2010, June 3). Eyes of cattle may become new windows to detect mad cow disease. ScienceDaily.

Thomas, F. P. (2010). Variant Creutzfeldt-Jakob Disease and Bovine Spongiform Encephalopathy.

Ummat A, D. A. Nanorobotics.

Utah,   U.   o.   (2010).   Retrieved   from http://learn.genetics.utah.edu/content/begin/dna/prions/controversy.html

Vass, D. A. (2001). Beyond the grave – understanding human decomposition. Microbiology Today .

Acknowledgments

This book would not have been possible without the support of excellent people and amazing organisations. We apologise in advance to anybody we miss from this list. If you want to come and argue about it with us remember we have our own crossbows.

University of Glasgow, the Wellcome Trust, Doctor Russell, Doctor Doug, Doctor Kevin, Doctor Katie, Doctor Duncan, Doctor Alison, Doctor Peter, Doctor Pete, Doctor Heff, Doctor Green, Doctor Suzanne, Doctor Colin, Doctor Drew, Doctor Sean, Doctor Gary, and Professor Sparks.

Lightning Source UK Ltd.
Milton Keynes UK
04 April 2011

170243UK00004B/1/P